INVESTIGATION OF
SITE CHARACTERISTICS AND
EVALUATION OF RADIATION
RISKS TO THE PUBLIC
AND THE ENVIRONMENT
IN SITE EVALUATION
FOR NUCLEAR INSTALLATIONS

The following States are Members of the International Atomic Energy Agency:

AFGHANISTAN	GEORGIA	PAKISTAN
ALBANIA	GERMANY	PALAU
ALGERIA	GHANA	PANAMA
ANGOLA	GREECE	PAPUA NEW GUINEA
ANTIGUA AND BARBUDA	GRENADA	PARAGUAY
ARGENTINA	GUATEMALA	PERU
ARMENIA	GUINEA	PHILIPPINES
AUSTRALIA	GUYANA	POLAND
AUSTRIA	HAITI	PORTUGAL
AZERBAIJAN	HOLY SEE	QATAR
BAHAMAS, THE	HONDURAS	REPUBLIC OF MOLDOVA
BAHRAIN	HUNGARY	ROMANIA
BANGLADESH	ICELAND	RUSSIAN FEDERATION
BARBADOS	INDIA	RWANDA
BELARUS	INDONESIA	SAINT KITTS AND NEVIS
BELGIUM	IRAN, ISLAMIC REPUBLIC OF	SAINT LUCIA
BELIZE	IRAQ	SAINT VINCENT AND
BENIN	IRELAND	THE GRENADINES
BOLIVIA, PLURINATIONAL	ISRAEL	SAMOA
STATE OF	ITALY	SAN MARINO
BOSNIA AND HERZEGOVINA	JAMAICA	SAUDI ARABIA
BOTSWANA	JAPAN	SENEGAL
BRAZIL	JORDAN	SERBIA
BRUNEI DARUSSALAM	KAZAKHSTAN	SEYCHELLES
BULGARIA	KENYA	SIERRA LEONE
BURKINA FASO	KOREA, REPUBLIC OF	SINGAPORE
BURUNDI	KUWAIT	SLOVAKIA
CABO VERDE	KYRGYZSTAN	SLOVENIA
CAMBODIA	LAO PEOPLE'S DEMOCRATIC	SOMALIA
CAMEROON	REPUBLIC	SOUTH AFRICA
CANADA	LATVIA	SPAIN
CENTRAL AFRICAN	LEBANON	SRI LANKA
REPUBLIC	LESOTHO	SUDAN
CHAD	LIBERIA	SWEDEN
CHILE	LIBYA	SWITZERLAND
CHINA	LIECHTENSTEIN	SYRIAN ARAB REPUBLIC
COLOMBIA	LITHUANIA	TAJIKISTAN
COMOROS	LUXEMBOURG	THAILAND
CONGO	MADAGASCAR	TOGO
COOK ISLANDS	MALAWI	TONGA
COSTA RICA	MALAYSIA	TRINIDAD AND TOBAGO
CÔTE D'IVOIRE	MALI	TUNISIA
CROATIA	MALTA	TÜRKİYE
CUBA	MARSHALL ISLANDS	TURKMENISTAN
CYPRUS	MAURITANIA	UGANDA
CZECH REPUBLIC	MAURITIUS	UKRAINE
DEMOCRATIC REPUBLIC	MEXICO	UNITED ARAB EMIRATES
OF THE CONGO	MONACO	UNITED KINGDOM OF
DENMARK	MONGOLIA	GREAT BRITAIN AND
DJIBOUTI	MONTENEGRO	NORTHERN IRELAND
DOMINICA	MOROCCO	UNITED REPUBLIC OF TANZANIA
DOMINICAN REPUBLIC	MOZAMBIQUE	UNITED STATES OF AMERICA
ECUADOR	MYANMAR	URUGUAY
EGYPT	NAMIBIA	UZBEKISTAN
EL SALVADOR	NEPAL	VANUATU
ERITREA	NETHERLANDS,	VENEZUELA, BOLIVARIAN
ESTONIA	KINGDOM OF THE	REPUBLIC OF
ESWATINI	NEW ZEALAND	VIET NAM
ETHIOPIA	NICARAGUA	YEMEN
FIJI	NIGER	ZAMBIA
FINLAND	NIGERIA	ZIMBABWE
FRANCE	NORTH MACEDONIA	
GABON	NORWAY	
GAMBIA, THE	OMAN	

IAEA SAFETY STANDARDS SERIES No. SSG-92

INVESTIGATION OF SITE CHARACTERISTICS AND EVALUATION OF RADIATION RISKS TO THE PUBLIC AND THE ENVIRONMENT IN SITE EVALUATION FOR NUCLEAR INSTALLATIONS

SPECIFIC SAFETY GUIDE

INTERNATIONAL ATOMIC ENERGY AGENCY
VIENNA, 2025

IAEA Library Cataloguing in Publication Data

Names: International Atomic Energy Agency.
Title: Investigation of site characteristics and evaluation of radiation risks to the public and the environment in site evaluation for nuclear installations / International Atomic Energy Agency.
Description: Vienna : International Atomic Energy Agency, 2025. | Series: Safety standards series, ISSN 1020-525X ; no. SSG-92 | Includes bibliographical references.
Identifiers: IAEAL 25-01796 | ISBN 978-92-0-116425-4 (paperback : alk. paper) | ISBN 978-92-0-116525-1 (pdf) | ISBN 978-92-0-116625-8 (epub)
Subjects: LCSH: Nuclear facilities — Safety measures. | Nuclear facilities — Environmental aspects. | Radiation — Health aspects.
Classification: UDC 614.876 | STI/PUB/2122

FOREWORD

by Rafael Mariano Grossi
Director General

The IAEA's Statute authorizes it to "establish...standards of safety for protection of health and minimization of danger to life and property". These are standards that the IAEA must apply to its own operations, and that States can apply through their national regulations.

The IAEA started its safety standards programme in 1958 and there have been many developments since. As Director General, I am committed to ensuring that the IAEA maintains and improves upon this integrated, comprehensive and consistent set of up to date, user friendly and fit for purpose safety standards of high quality. Their proper application in the use of nuclear science and technology should offer a high level of protection for people and the environment across the world and provide the confidence necessary to allow for the ongoing use of nuclear technology for the benefit of all.

Safety is a national responsibility underpinned by a number of international conventions. The IAEA safety standards form a basis for these legal instruments and serve as a global reference to help parties meet their obligations. While safety standards are not legally binding on Member States, they are widely applied. They have become an indispensable reference point and a common denominator for the vast majority of Member States that have adopted these standards for use in national regulations to enhance safety in nuclear power generation, research reactors and fuel cycle facilities as well as in nuclear applications in medicine, industry, agriculture and research.

The IAEA safety standards are based on the practical experience of its Member States and produced through international consensus. The involvement of the members of the Safety Standards Committees, the Nuclear Security Guidance Committee and the Commission on Safety Standards is particularly important, and I am grateful to all those who contribute their knowledge and expertise to this endeavour.

The IAEA also uses these safety standards when it assists Member States through its review missions and advisory services. This helps Member States in the application of the standards and enables valuable experience and insight to be shared. Feedback from these missions and services, and lessons identified from events and experience in the use and application of the safety standards, are taken into account during their periodic revision.

I believe the IAEA safety standards and their application make an invaluable contribution to ensuring a high level of safety in the use of nuclear technology. I encourage all Member States to promote and apply these standards, and to work with the IAEA to uphold their quality now and in the future.

THE IAEA SAFETY STANDARDS

BACKGROUND

Radioactivity is a natural phenomenon and natural sources of radiation are features of the environment. Radiation and radioactive substances have many beneficial applications, ranging from power generation to uses in medicine, industry and agriculture. The radiation risks to workers and the public and to the environment that may arise from these applications have to be assessed and, if necessary, controlled.

Activities such as the medical uses of radiation, the operation of nuclear installations, the production, transport and use of radioactive material, and the management of radioactive waste must therefore be subject to standards of safety.

Regulating safety is a national responsibility. However, radiation risks may transcend national borders, and international cooperation serves to promote and enhance safety globally by exchanging experience and by improving capabilities to control hazards, to prevent accidents, to respond to emergencies and to mitigate any harmful consequences.

States have an obligation of diligence and duty of care, and are expected to fulfil their national and international undertakings and obligations.

International safety standards provide support for States in meeting their obligations under general principles of international law, such as those relating to environmental protection. International safety standards also promote and assure confidence in safety and facilitate international commerce and trade.

A global nuclear safety regime is in place and is being continuously improved. IAEA safety standards, which support the implementation of binding international instruments and national safety infrastructures, are a cornerstone of this global regime. The IAEA safety standards constitute a useful tool for contracting parties to assess their performance under these international conventions.

THE IAEA SAFETY STANDARDS

The status of the IAEA safety standards derives from the IAEA's Statute, which authorizes the IAEA to establish or adopt, in consultation and, where appropriate, in collaboration with the competent organs of the United Nations and with the specialized agencies concerned, standards of safety for protection of health and minimization of danger to life and property, and to provide for their application.

With a view to ensuring the protection of people and the environment from harmful effects of ionizing radiation, the IAEA safety standards establish fundamental safety principles, requirements and measures to control the radiation exposure of people and the release of radioactive material to the environment, to restrict the likelihood of events that might lead to a loss of control over a nuclear reactor core, nuclear chain reaction, radioactive source or any other source of radiation, and to mitigate the consequences of such events if they were to occur. The standards apply to facilities and activities that give rise to radiation risks, including nuclear installations, the use of radiation and radioactive sources, the transport of radioactive material and the management of radioactive waste.

Safety measures and security measures[1] have in common the aim of protecting human life and health and the environment. Safety measures and security measures must be designed and implemented in an integrated manner so that security measures do not compromise safety and safety measures do not compromise security.

The IAEA safety standards reflect an international consensus on what constitutes a high level of safety for protecting people and the environment from harmful effects of ionizing radiation. They are issued in the IAEA Safety Standards Series, which has three categories (see Fig. 1).

Safety Fundamentals

Safety Fundamentals present the fundamental safety objective and principles of protection and safety, and provide the basis for the Safety Requirements. The principles are expressed as 'must' statements.

Safety Requirements

Safety Requirements are governed by the objective and principles of the Safety Fundamentals. They establish the requirements to be met to ensure the protection of people and the environment, both now and in the future. The format and style of the Safety Requirements facilitate their use for the establishment of a national regulatory framework. Requirements are presented as 'overarching' requirements[2] in bold, followed by a number of associated requirements; all are equally important and are expressed as 'shall' statements.

Safety Guides

Safety Guides provide recommendations on how to comply with the Safety Requirements, indicating an international consensus that it is necessary to take the

---

[1]   See also publications issued in the IAEA Nuclear Security Series.

[2]   The IAEA Regulations for the Safe Transport of Radioactive Material do not include overarching requirements.

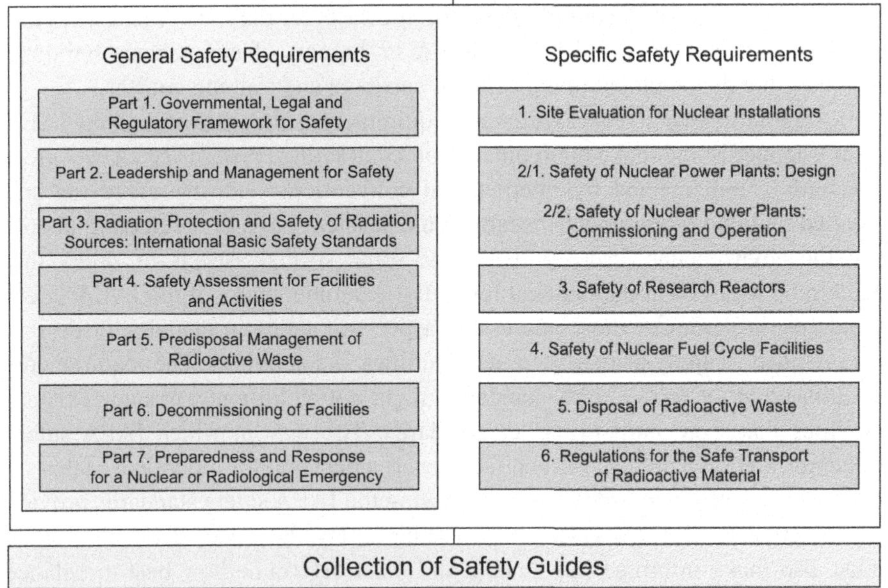

FIG. 1. The long term structure of the IAEA Safety Standards Series.

measures recommended (or alternative measures that achieve the same level of protection). Safety Guides present international good practices and, increasingly, best practices. The recommendations provided in Safety Guides are expressed as 'should' statements.

APPLICATION OF THE IAEA SAFETY STANDARDS

The principal users of safety standards in IAEA Member States are regulatory bodies and other relevant national authorities. The IAEA safety standards are also used by co-sponsoring organizations and by many organizations that design, construct and operate nuclear facilities, as well as organizations involved in the use of radiation and radioactive sources.

The IAEA safety standards are applicable, as relevant, throughout the entire lifetime of all facilities and activities — existing and new — utilized for peaceful purposes and to protective actions to reduce existing radiation risks. They can be used by States as a reference for their national regulations in respect of facilities and activities.

The IAEA's Statute makes the safety standards binding on the IAEA in relation to its own operations and also on States in relation to IAEA assisted operations.

The IAEA safety standards also form the basis for the IAEA's safety review services, and they are used by the IAEA in support of competence building, including the development of educational curricula and training courses.

International conventions contain requirements similar to those in the IAEA safety standards and make them binding on contracting parties. The IAEA safety standards, supplemented by international conventions, industry standards and detailed national requirements, establish a consistent basis for protecting people and the environment. There will also be some special aspects of safety that need to be assessed at the national level. For example, many of the IAEA safety standards, in particular those addressing aspects of safety in planning or design, are intended to apply primarily to new facilities and activities. The requirements established in the IAEA safety standards might not be fully met at some existing facilities that were built to earlier standards. The way in which IAEA safety standards are to be applied to such facilities is a decision for individual States.

The scientific considerations underlying the IAEA safety standards provide an objective basis for decisions concerning safety; however, decision makers must also make informed judgements and must determine how best to balance the benefits of an action or an activity against the associated radiation risks and any other detrimental impacts to which it gives rise.

DEVELOPMENT PROCESS FOR THE IAEA SAFETY STANDARDS

The preparation and review of the safety standards involves the IAEA Secretariat and five Safety Standards Committees, for emergency preparedness and response (EPReSC) (as of 2016), nuclear safety (NUSSC), radiation safety (RASSC), the safety of radioactive waste (WASSC) and the safe transport of radioactive material (TRANSSC), and a Commission on Safety Standards (CSS) which oversees the IAEA safety standards programme (see Fig. 2).

All IAEA Member States may nominate experts for the Safety Standards Committees and may provide comments on draft standards. The membership of the Commission on Safety Standards is appointed by the Director General and includes senior governmental officials having responsibility for establishing national standards.

A management system has been established for the processes of planning, developing, reviewing, revising and establishing the IAEA safety standards. It articulates the mandate of the IAEA, the vision for the future application of the safety standards, policies and strategies, and corresponding functions and responsibilities.

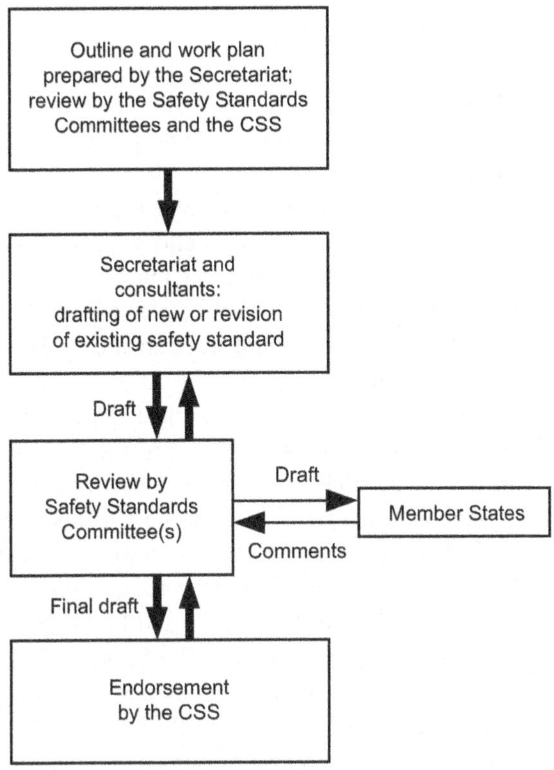

FIG. 2. *The process for developing a new safety standard or revising an existing standard.*

INTERACTION WITH OTHER INTERNATIONAL ORGANIZATIONS

The findings of the United Nations Scientific Committee on the Effects of Atomic Radiation (UNSCEAR) and the recommendations of international expert bodies, notably the International Commission on Radiological Protection (ICRP), are taken into account in developing the IAEA safety standards. Some safety standards are developed in cooperation with other bodies in the United Nations system or other specialized agencies, including the Food and Agriculture Organization of the United Nations, the United Nations Environment Programme, the International Labour Organization, the OECD Nuclear Energy Agency, the Pan American Health Organization and the World Health Organization.

INTERPRETATION OF THE TEXT

Safety related terms are to be understood as they appear in the IAEA Nuclear Safety and Security Glossary (see https://www.iaea.org/resources/publications/iaea-nuclear-safety-and-security-glossary). Otherwise, words are used with the spellings and meanings assigned to them in the latest edition of The Concise Oxford Dictionary. For Safety Guides, the English version of the text is the authoritative version.

The background and context of each standard in the IAEA Safety Standards Series and its objective, scope and structure are explained in Section 1, Introduction, of each publication.

Material for which there is no appropriate place in the body text (e.g. material that is subsidiary to or separate from the body text, is included in support of statements in the body text, or describes methods of calculation, procedures or limits and conditions) may be presented in appendices or annexes.

An appendix, if included, is considered to form an integral part of the safety standard. Material in an appendix has the same status as the body text, and the IAEA assumes authorship of it. Annexes and footnotes to the main text, if included, are used to provide practical examples or additional information or explanation. Annexes and footnotes are not integral parts of the main text. Annex material published by the IAEA is not necessarily issued under its authorship; material under other authorship may be presented in annexes to the safety standards. Extraneous material presented in annexes is excerpted and adapted as necessary to be generally useful.

CONTENTS

1. INTRODUCTION

BACKGROUND

1.1. IAEA Safety Standards Series No. SSR-1, Site Evaluation for Nuclear Installations [1], establishes requirements for:

(a) Defining the information to be used in the site evaluation process;
(b) Evaluating a site such that the site specific hazards and the safety related site characteristics are adequately taken into account, in order to derive appropriate site specific design parameters;
(c) Investigating site characteristics and assessing the radiological environmental impact of nuclear installations;
(d) Analysing the characteristics of the population and the region surrounding the site to determine whether there would be significant difficulties in implementing emergency response actions effectively.

1.2. IAEA Safety Standards Series Nos SSR-3, Safety of Research Reactors [2], SSR-4, Safety of Nuclear Fuel Cycle Facilities [3], GSR Part 3, Radiation Protection and Safety of Radiation Sources: International Basic Safety Standards [4], and GSR Part 7, Preparedness and Response for a Nuclear or Radiological Emergency [5], establish requirements on these topics. This Safety Guide provides recommendations on how to meet these requirements.

1.3. This Safety Guide takes into account progress in the investigation of site characteristics and assessment of the radiological environmental impact of nuclear installations, as well as in regulatory practices in Member States. It considers lessons identified from discharges[1] and accident conditions at nuclear installations, feedback from safety review missions and the results of recent research in relevant fields.

1.4. This Safety Guide provides new or updated recommendations that address the following topics:

(a) Recent updates in the investigation of site characteristics and radiological environmental impact assessment for nuclear installations;

---

[1] As defined in the IAEA Nuclear Safety and Security Glossary [6], a discharge is a planned and controlled release of (usually gaseous or liquid) radioactive substances to the environment.

(b)   Methodologies for the analysis of dispersion and transfer of radionuclides;
(c)   Linking of the results of analyses with the assessment of overall radiological impact (including dose assessment);
(d)   Development of a complete set of potential release scenarios;
(e)   Application of a graded approach to the radiological environmental impact assessment of nuclear installations;
(f)   Monitoring of radioactivity in the environment.

1.5.   This Safety Guide supersedes IAEA Safety Standards Series No. NS-G-3.2, Dispersion of Radioactive Material in Air and Water and Consideration of Population Distribution in Site Evaluation for Nuclear Power Plants.[2]

OBJECTIVE

1.6.   The main objective of this Safety Guide is to provide recommendations on the investigation of site characteristics and the evaluation of radiation risks[3] to the public and the environment in site evaluation for nuclear installations, to meet the applicable safety requirements established in SSR-1 [1], SSR-3 [2], SSR-4 [3], GSR Part 3 [4] and GSR Part 7 [5]. It provides specific recommendations on radiological environmental impact assessment for nuclear installations in the site evaluation process.

1.7.   This Safety Guide is intended for use by organizations involved in the investigation of site characteristics and the evaluation of radiation risks for nuclear installations, which includes radiological environmental impact assessment (e.g. by operating organizations, designers or technical support organizations), as well as by regulatory bodies. The applicability of the guidance will vary depending on site characteristics and the life stage of a nuclear installation.

---

[2] INTERNATIONAL ATOMIC ENERGY AGENCY, Dispersion of Radioactive Material in Air and Water and Consideration of Population Distribution in Site Evaluation for Nuclear Power Plants, IAEA Safety Standards Series No. NS-G-3.2, IAEA, Vienna (2002).

[3] As defined in the IAEA Nuclear Safety and Security Glossary [6], radiation risks are detrimental health effects of exposure to radiation (including the likelihood of such effects occurring), and any other safety related risks (including those to the environment) that might arise as a direct consequence of: (a) Exposure to radiation; (b) The presence of radioactive material (including radioactive waste) or its release to the environment; (c) A loss of control over a nuclear reactor core, nuclear chain reaction, radioactive source or any other source of radiation.

SCOPE

1.8. This Safety Guide provides recommendations on how to assess the radiological environmental impact due to discharges and accidental releases from a new or existing nuclear installation on people and the environment. It covers the investigation of site characteristics, such as population distribution, land and water use in the region, background levels of radioactivity in the environmental media, and meteorological, hydrological and hydrogeological characteristics of the region.

1.9. This Safety Guide provides recommendations on the development of the site evaluation report for a nuclear installation. It also provides recommendations on the development of the radiological impact assessment, which is part of the environmental impact assessment report, and relevant sections of the safety analysis report. The recommendations apply to the site evaluation process and other assessments performed throughout the lifetime of a nuclear installation (e.g. periodic safety review) or when site characteristics change.

1.10. This Safety Guide also covers analysis of the dispersion of radionuclides in the atmosphere, surface water and groundwater, and assessment of overall radiological impact and dose assessment in the process of site evaluation.

1.11. Data collection within the scope of this publication (e.g. site specific natural and infrastructural conditions in the region of the site) supports the determination of off-site emergency planning zones and emergency planning distances. The feasibility of planning effective emergency response actions and the application of a management system for those actions are also addressed.

1.12. Although an environmental impact assessment covers potential radiological and non-radiological impacts, the latter are out of the scope of this Safety Guide. Nevertheless, care should be taken to ensure that the non-radiological impact assessment is consistent with the radiological impact assessment in terms of estimating the transfer of releases to the environment and defining the representative person.

1.13. Environmental impacts of alternatives, which need to be considered as part of the environmental impact assessment, are not a subject of this Safety Guide. The conditions at the site and in the vicinity could change over time (e.g. population size, distribution) and this needs to be assessed. However, this Safety Guide covers present conditions (e.g. current nuclear facilities on the site) as part of the assessment of environmental impacts from the proposed actions.

1.14. In addition to the characteristics investigated in this Safety Guide, other site characteristics are considered in site evaluation for a nuclear installation. They relate to natural external hazards and include meteorological hazards (e.g. lightning, extreme winds, tropical cyclones, typhoons and hurricanes, tornadoes, waterspouts, extreme rainfall and snowfall, extreme temperatures), hydrological hazards (e.g. storm surges, waves, tsunamis, seiches), seismic hazards (e.g. vibratory ground motion hazards, fault displacements), geological hazards (e.g. slope instability, collapse, subsidence or uplift, soil liquefaction) and volcanic hazards (e.g. tephra fallout, pyroclastic flows, lava flows). Site characteristics also relate to hazards associated with human induced events (e.g. hazards associated with events occurring on nearby land, rivers or seas or in the air due to collisions and explosions, fire, missile generation and releases of hazardous gases from industrial facilities near the site). These characteristics are covered in more detail in IAEA Safety Standards Series Nos SSG-18, Meteorological and Hydrological Hazards in Site Evaluation for Nuclear Installations [7]; SSG-9 (Rev. 1), Seismic Hazards in Site Evaluation for Nuclear Installations [8]; SSG-93, Geotechnical Aspects in the Siting and Design of Nuclear Installations [9]; SSG-21, Volcanic Hazards in Site Evaluation for Nuclear Installations [10]; and SSG-79, Hazards Associated with Human Induced External Events in Site Evaluation for Nuclear Installations [11]. Any design changes implemented as a result of the external hazard assessment will affect the radiological impact assessments for the nuclear installation.

1.15. The recommendations provided in this Safety Guide are applicable to all types of nuclear installations as defined in Ref. [6], but not to transportable installations. Although the recommendations are predominantly written with nuclear power plants in mind, they are also applicable to other nuclear installations through the use of a graded approach (see Section 10).

1.16. The assessments of radiological environmental impacts due to malicious acts at nuclear installations are outside the scope of this Safety Guide, although some of the recommendations provided might also be consistent with the needs of nuclear security. Specific guidance on the protection of nuclear power plants against malicious acts is provided in IAEA Nuclear Security Series No. 48-T, Identification and Categorization of Sabotage Targets, and Identification of Vital Areas at Nuclear Facilities [12].

STRUCTURE

1.17. Section 2 summarizes the relevant safety requirements and provides recommendations on identifying exposure pathways, the general approach to environmental impact assessment, site characterization for such an assessment, the assessment of transfer of radionuclides to the environment, the approach for existing sites and addressing climate change. Recommendations on baseline environmental characteristics of the site and region, including population distribution, are provided in Section 3. Recommendations on the analysis of the dispersion of radionuclides in the atmosphere are provided in Section 4. Recommendations on the analysis of the dispersion of radionuclides in surface water and groundwater are provided in Sections 5 and 6, respectively. Recommendations on the assessment of overall radiological impact are provided in Section 7. Section 8 provides recommendations on the monitoring of radioactivity in the environment. Recommendations in relation to the feasibility of effective emergency response actions are provided in Section 9. Recommendations on the application of a graded approach to radiological environmental impact assessment for nuclear installations are provided in Section 10. Section 11 provides recommendations on the application of the management system for activities that are performed for the investigation of site characteristics and evaluation of radiation risks to the public and the environment in site evaluation for nuclear installations. The Appendix provides a methodology for the application of a graded approach to the analysis of dispersion of radionuclides in groundwater.

2. REQUIREMENTS FOR AND THE GENERAL APPROACH TO INVESTIGATING SITE CHARACTERISTICS AND EVALUATING RADIATION RISKS IN SITE EVALUATION FOR NUCLEAR INSTALLATIONS

REQUIREMENTS FOR EVALUATION OF THE POTENTIAL EFFECTS OF NUCLEAR INSTALLATIONS ON PEOPLE AND THE ENVIRONMENT

2.1. Requirement 5 of SSR-1 [1] states:

"The site and the region shall be investigated with regard to the characteristics that could affect the safety of the nuclear installation and the potential radiological impact of the nuclear installation on people and the environment."

2.2. Requirement 12 of SSR-1 [1] states:

"In determining the potential radiological impact of the nuclear installation on the region for operational states and accident conditions, including accidents that could warrant emergency response actions, appropriate estimates shall be made of the potential releases of radioactive material, with account taken of the design of the nuclear installation and its safety features."

2.3. Requirement 25 of SSR-1 [1] states that **"The dispersion in air and water of radioactive material released from the nuclear installation in operational states and in accident conditions shall be assessed."**

2.4. Requirement 29 of SSR-1 [1] states:

"All natural and human induced external hazards and site conditions shall be periodically reviewed by the operating organization as part of the periodic safety review and as appropriate throughout the lifetime of the nuclear installation, with due account taken of operating experience and new safety related information."

2.5. Paragraph 7.5 of SSR-1 [1] states:

"The site specific external hazards and the site conditions shall be re-evaluated, as necessary, based on the outcome of the periodic review of site specific hazards or because of new data relevant to the radiological environmental impact assessment or to the safe operation of the nuclear installation."

2.6. Paragraphs 5.1–5.12 of SSR-3 [2] provide requirements for site evaluation specific to research reactors.

2.7. Paragraphs 5.1–5.12 of SSR-4 [3] provide requirements for site evaluation specific to nuclear fuel cycle facilities.

IDENTIFICATION OF EXPOSURE PATHWAYS IN SITE EVALUATION FOR NUCLEAR INSTALLATIONS

2.8. Paragraph 4.39 of SSR-1 [1] states:

"The direct and indirect pathways by which radioactive releases from the nuclear installation could potentially affect the public and the environment shall be identified and evaluated. In this evaluation, specific regional and site characteristics, including the population distribution in the region, shall be taken into account, with special attention paid to the transport and accumulation of radionuclides in the biosphere."

2.9. Exposure pathways are routes by which radiation or radionuclides can reach humans and cause exposure. Possible exposure pathways for radionuclides released to the atmosphere and surface water during normal operation of nuclear installations such as nuclear power plants are listed in para. 5.27 of IAEA Safety Standards Series No. GSG-10, Prospective Radiological Environmental Impact Assessment for Facilities and Activities [13]. Figure 1 of IAEA Safety Standards Series No. GSG-19, Monitoring for Protection of the Public and the Environment [14], shows the possible exposure pathways for members of the public as a result of releases of radioactive substances to the environment. The exposure pathways are as follows:

(a) Inhalation of airborne material in an atmospheric plume (e.g. gases, vapours, aerosols);
(b) Inhalation of resuspended material;
(c) Ingestion of crops;

(d)   Ingestion of animal food products (e.g. milk, meat, eggs);

(e)   Ingestion of drinking water;

(f)   Ingestion of aquatic food (e.g. freshwater or seawater fish, crustaceans, molluscs);

(g)   Ingestion of forest food (e.g. wild mushrooms, wild berries, game);

(h)   Ingestion of breast milk or locally prepared food for infants;

(i)   Inadvertent ingestion of soil and sediments;

(j)   External exposure from radionuclides in an atmospheric plume (i.e. cloud shine);

(k)   External exposure from radionuclides deposited on the ground (i.e. ground shine) and on surfaces;

(l)   External exposure from radionuclides in water and sediments (e.g. from activities on shores, swimming and fishing).

2.10. In addition, there might be exposure due to irradiation from activity deposited from the atmospheric plume on people's skin, clothing or vehicles, and exposure due to irradiation from activity on the site (see the additional exposure pathways described in para. 5.29 of GSG-10 [13]) without any activity having been released from the installation. As these would also be expected pathways for accidental releases, their relative significance is much higher during such situations.

2.11. The first step in evaluating a site for a nuclear installation in terms of radiation risks should be to identify all possible exposure pathways and then determine which are significant in terms of exposure of the public or the environment.

2.12. The significance of the pathway depends on the quantities and the chemical and physical form of the radionuclides released. It also depends on other characteristics of the release that might affect the subsequent dispersion of radionuclides and their behaviour in the environment, the location and medium into which the release is made and the characteristics of the environment and population around the site.

2.13. It should therefore be confirmed that significant pathways have been identified, especially if there are specific characteristics about the installation design, its operation, the site, land and water use around the site, farming practices or the surrounding location. For example, the presence of desalination plants producing water (either at or close to the discharge outlets) for drinking or irrigation might give rise to exposure pathways.

2.14. Accidental releases may encompass a spectrum of scenarios, such as design basis accidents and design extension conditions, including sequences without significant fuel degradation and sequences with core melting, involving different source terms and release locations and releases to different media. Different States may use different criteria to assess the adverse consequences from accidental releases, such as individual dose, or a risk measure (i.e. a combination of a consequence and the likelihood of that consequence occurring) such as individual risk or societal risk. In most cases, atmospheric releases are the dominant contributors to the total risk but there might be other scenarios involving accidental releases to surface water or groundwater. Evaluating the consequences of accidental releases to surface water and groundwater may necessitate detailed analysis. However, if the conditional probability of a consequence is calculated, with a reasonable level of confidence that is considered acceptable to the regulatory body, and it does not make a significant contribution to the overall risk, then detailed analysis of the consequence might not be needed. If both the probability of occurrence and the estimated contribution of a potential hazard to the overall risk are significant, then a detailed analysis should be performed. Reference [15] presents criteria for assessing overall risk.

2.15. Once the significant exposure pathways have been identified, the local environment should be characterized sufficiently to allow exposure from these pathways to be calculated with an appropriate level of detail. The level of characterization detail should be commensurate with the importance of the pathway for the particular scenario being modelled. In accordance with para. 4.1 of SSR-1 [1], a graded approach, commensurate with the radiation risk posed to people and the environment, is required to be applied for this purpose (see also Section 10).

2.16. During operational states (i.e. normal operation and anticipated operational occurrences), there are authorized and regulated effluent discharges to the atmosphere and surface water. In accident conditions, additionally there might be direct releases to groundwater or to the ground surface. The initial release into each of these media and the resulting significant exposure pathways are discussed in paras 2.17–2.31. However, as a result of facility design, there should be no, or only minor, radiological impact beyond the immediate vicinity of the installation as a result of anticipated operational occurrences or design basis accidents (see para. 7.31 of IAEA Safety Standards Series No. SSG-2 (Rev. 1), Deterministic Safety Analysis for Nuclear Power Plants [16]).

Atmospheric releases

2.17. The significant exposure pathways depend on the nature of the atmospheric release, including the source term, location, and effective height at which the release is made. Recommendations on the determination of the source term for releases to the environment for anticipated operational occurrences and accident conditions are provided in SSG-2 (Rev. 1) [16]. The source term defines the quantities and the physical, isotopic and chemical forms of the radionuclides released, as well as the time profile of the release. Factors affecting the subsequent transfer and behaviour of the radionuclides in the environment include the stack height and the energy associated with the atmospheric release.

2.18. For discharges under normal operation, the measures taken to mitigate the atmospheric release, control the discharge and ensure that exposures are as low as reasonably achievable (taking into account economic and social factors; see GSR Part 3 [4]) and in compliance with regulatory and operational limits, tend to focus on radionuclides and pathways that are radiologically significant. For this reason, some less obvious radionuclides (e.g. radiocarbon (^{14}C), tritium (^{3}H)), which can be difficult to remove, and those that might accumulate in the environment during the lifetime of the installation and/or less obvious pathways might become more significant. Additionally, facility design should ensure negligible radiological impact beyond the immediate vicinity of the plant from any anticipated operational occurrence. The radiological acceptance criteria for doses and correspondingly for releases for each anticipated operational occurrence should be comparable with annual limits for normal operation and more restrictive than for design basis accidents. Acceptable effective dose limits are similar to those for normal operation (see para. 7.23 of SSG-2 (Rev. 1) [16]).

2.19. The significant exposure pathways from atmospheric releases can also be identified through monitoring of the environment (see Section 8). Ideally, this monitoring should be performed for as long as there is a possibility of radioactive release to the environment (this is usually until the installation has been decommissioned). This allows any periodic (e.g. seasonal) or long term trends to be observed.

2.20. Discharges from nuclear installations are expected to continue throughout the lifetime of the nuclear installations, from construction to decommissioning, and therefore accumulation of activity in the environment over this period should be considered for those radionuclides whose half-lives are significant compared with the operating lifetime of the installation in question.

2.21. Radiocarbon and tritium can be particularly difficult to model in the environment because, whatever their chemical form at release, they can soon be incorporated into CO_2 or water, respectively, or be incorporated into organic molecules in environmental media and become part of the food chain, contributing in this way to individual dose.

2.22. For an accidental atmospheric release leading to exposure of the public, the significant pathways depend on the source term and the nature of the release. The most common scenario for an accidental atmospheric release is usually direct inhalation of the plume, which forms the most significant pathway in the short term as it is a direct route of internal exposure of radionuclides in gas or vapour forms or as suspended particulate. An atmospheric release scenario in which inhalation would not be a significant pathway would involve only noble gases (which are not absorbed by the body when inhaled) or a scenario where the initial plume does not lead directly to exposure of people.

2.23. Other pathways that can lead to exposure are ground shine (i.e. radiation from activity deposited on the ground), sky shine (i.e. radiation deflected by the air) and cloud shine (i.e. radiation from activity in an airborne plume). These pathways are usually less significant than direct inhalation for members of the public.

2.24. For the pathways listed in para. 2.9(b)–(i), (k) and (l), exposure usually occurs as a result of deposition of radionuclides from the plume. Deposition can be either 'dry' or 'wet'. Dry deposition occurs when contaminants in the plume adsorb to suspended particulates in the air, which are then deposited on the ground (for an elevated release, deposition might be some distance from the release point). Wet deposition occurs when precipitation (e.g. rain, snow) washes material from the plume. Dry or wet deposition leads to plume depletion.

2.25. If deposition occurs, the first pathway that should be considered is direct radiation from the deposited activity. The significance of this pathway depends on the rate of deposition, which in turn depends on the chemical and physical form of the nuclides and their radiation emissions. For example, elemental iodine as a reactive chemical form has a high deposition velocity. The relative significance of different radionuclides also depends on the time frame over which the dose is integrated: radionuclides with longer radioactive half-lives and longer biological half-lives become increasingly important for longer integration times. For integration times comparable with the duration of the direct inhalation of the plume (typically a few days), inhalation is usually the dominant pathway; for much longer integration times (e.g. years), the deposition pathway can become dominant for longer lived radionuclides. External exposure from deposition can be a long term pathway of exposure for long lived radionuclides to members of the public. For long integration times (e.g. > 1 year), the weathering and migration of radionuclides through soils — which reduce the dose from this pathway — may also need to be considered.

2.26. The quantities of radionuclides deposited on the ground are also important in determining the dose by ingestion. Compared with direct inhalation, the impact is usually lower since only a small fraction of the plume is deposited and incorporated into the food chain and there is some time delay before consumption, during which short lived radionuclides decay. Contamination and consumption of crops is a pathway that can lead to exposure far from the release point as the produce is transported. However, monitoring of commercially produced food, milk and drinking water and the application of operational intervention levels (see GSR Part 7 [5]) should lead to the control of exposure via this pathway. Ingestion of forest food (e.g. wild mushrooms, wild berries, game) might be less amenable to control, but its impact is usually limited to areas close to the original contamination. This pathway can be a significant contributor to the risk of exposure for individuals and hence to societal risk (e.g. total number of fatalities in the exposed population, total economic cost, societal stress) [17].

2.27. Resuspension of deposited radionuclides that are then inhaled leads to a longer term impact on the public, but given that only a small fraction of the plume is deposited and then resuspended, the impact on any individual is negligible compared with direct inhalation. For those people who do not inhale the plume directly during an accidental release, resuspension should be considered as a possibly significant pathway.

Discharges and releases to surface water

2.28. Discharges and releases to surface water can lead to human exposure and exposure of non-human species due to direct irradiation from activity in the water and suspended sediments, although water provides much more shielding than air provides for atmospheric releases. Activity might also accumulate in sediments (which can lead to exposure pathways such as direct radiation and possible eventual resuspension and inhalation) as a result of releases to surface water. Radionuclides can also enter the aquatic food chain. These activities should also be considered when determining the relative significance of surface water pathways.

2.29. For accidental releases to surface water, the shielding provided by the water, the lower likelihood of anybody being directly exposed compared with an atmospheric release, and the greater dispersion (especially for releases to the sea) usually mean that aquatic pathways are less significant than terrestrial pathways.

These factors should be taken into account when determining the relative significance of surface water pathways.

Releases to groundwater

2.30. Direct releases to groundwater should not be authorized. However, radionuclides might enter groundwater indirectly, for example through exchange with river water in which discharges are allowed, via atmospheric release and subsequent deposition on the ground or surface water, or via sinking streams in karstic terrains. Accidental releases to groundwater might occur, for example as a result of spillage of radioactive waste, leaks of contaminated water, or core melt through the basemat for some nuclear installations. Unlike direct releases to the atmosphere, which lead to immediate exposure, activity released to groundwater might be transferred through the groundwater for many years before it reaches a location where exposure of the public could occur. During this time, short lived radionuclides decay, and the shielding of the ground significantly limits exposure by direct radiation. These accidental releases could, however, lead to long term contamination with few, if any, remediation solutions. Any extraction of groundwater for drinking water or irrigation could result in more immediate exposure via the pathways described in para. 6.37. However, this water could easily be monitored, and its use for drinking or irrigation could be suspended if activity levels were above the established limits. All these factors should be considered when determining the significance of this pathway.

Releases to ground surface

2.31. As with releases to groundwater (see para. 2.30), planned direct discharges to the ground surface should not be authorized. Accidental liquid releases to the ground surface could occur, however, leading to contaminated ground and potentially to resuspension of radionuclides or to radionuclides entering the food chain. This should be considered as a possibility when investigating different potential pathways.

GENERAL APPROACH TO RADIOLOGICAL ENVIRONMENTAL IMPACT ASSESSMENT FOR NUCLEAR INSTALLATIONS

2.32. The components of radiological environmental impact assessment for protection of the public and of non-human species in normal operation, and for consideration of potential exposure, are shown in figs 2 and 3 of GSG-10 [13].

2.33. The first step in conducting the assessment is to select the source term(s). The selection process might be complex, taking into account factors such as installation design, materials used and changes of inventory during operation of the installation (e.g. due to fission products). For installations using well known technology employed elsewhere, data from these other operations could be used to select and confirm the source term. For developments where the technology is yet to be decided, the plant parameter envelope approach could be taken initially, whereby the maximum source term for the options under consideration is used, based on published data from vendors or from previous projects. For small modular reactors, based on current technology, one option could be to scale the source terms from large reactors. For novel types of reactors (e.g. evolutionary and innovative designs), the only data available might be from the reactor vendors; in this case, the project developers should ensure that the source term estimates are conservative. For other types of installations, a similar or simplified approach may be adopted depending on the complexity and potential hazard associated with the installation. To the extent possible, in the light of the available data, several source terms should be selected corresponding to the different plant states (i.e. normal operation, anticipated operational occurrences, design basis accidents and design extension conditions, including severe accidents as defined in Ref. [6]) and considering the number of operating units to fully capture the environmental impact for normal and accident conditions (see paras 4.15–4.21). Further recommendations on selecting the source term(s) are provided in SSG-2 (Rev. 1) [16] and in IAEA Safety Standards Series No. SSG-4 (Rev. 1), Development and Application of Level 2 Probabilistic Safety Assessment for Nuclear Power Plants [18].

2.34. The next step is to model the dispersion of radionuclides in the environment (see Sections 4–6). The end points of the radiological environmental impact assessments are generally activity concentrations in the various environmental media (e.g. air, water, ground) that can lead to human exposure. These activity concentrations are then used to calculate the doses to a representative person (see Section 7). Exposure from direct radiation should also be calculated, if applicable.

2.35. Releases to the atmosphere or to water could lead to many people being exposed by several pathways. To assess the risk to the public, a representative person is selected, for whom the individual dose is calculated. Representative persons can be identified by determining the most significant exposure pathways (see para. 2.11). For example, for the inhalation pathway the representative person could be somebody living close to the site in the prevailing wind direction; for the ingestion pathway it could be a consumer who ingests a high fraction of locally produced food. However, all pathways need to be considered when assessing the dose to the representative person. Further recommendations on selecting the representative person are provided in Section 7, and guidance is also given in Ref. [19]. The doses from all the significant pathways that lead to exposure of the representative person should be added together to give the total effective dose.

SITE CHARACTERIZATION AND RADIOLOGICAL ENVIRONMENTAL IMPACT ASSESSMENT

2.36. The characterization of the environment should be sufficient to allow the radiation exposure of the public to be modelled for the purposes of the radiological environmental impact assessment. The characterization of the population, including their habits and land use, should be sufficient to allow the identification of a representative person. Further recommendations on baseline environmental data are provided in Section 3.

APPROACH TO ASSESSING THE TRANSFER OF RADIONUCLIDES TO THE ENVIRONMENT

2.37. For new sites, it is not necessary to model explicitly every single process involving the transfer of radioactivity between different environmental compartments. However, all processes should be considered and their relative significance assessed, allowing some processes to be discounted if their significance is small with regard to the impact on the end points being considered. If the effort involved, with regard to time and resources, would be disproportionate to the difference in the calculated end points — and considering other uncertainties, such as those in the source term — then few insights would be gained from detailed modelling. For example, for a postulated accidental release to the atmosphere, the uncertainties in the source terms might be larger than any differences in the end results caused by differences in the detailed modelling. Inevitably in these situations, some judgement may be involved, but any simplification made in the analysis should be justified. In making such judgements, the overall objectives of the analysis (e.g. demonstrating regulatory compliance) should be considered.

2.38. Sites with existing installations have the advantage that the environment should already have been characterized and there might be data from measured discharges and environmental monitoring that can inform the modelling of proposed discharges. There are two general cases in which new environmental impact assessments might be needed for an existing site:

(a)   Where a modification to an existing installation could affect the potential for releases;
(b)   Where a new installation is planned.

2.39. Where existing nuclear installations are located at or close to the site for the proposed new installation, the total impacts from the new installation in addition to those from existing installations should be considered.

CLIMATE CHANGE

2.40. Climate changes in terms of the meteorological, hydrological and geological conditions in the region of the nuclear installation site over the lifetime of the installation should be considered in the radiological environmental impact assessment. Due to the long term trends associated with climate change, the environmental impact assessment and the associated monitoring plans should be periodically reviewed and updated as necessary to reflect any changes that are identified and to take any necessary actions to reduce potential impacts to the environment.

3. BASELINE ENVIRONMENTAL DATA INCLUDING POPULATION DISTRIBUTION

3.1.   Requirement 3 of SSR-1 [1] states:

"The scope of the site evaluation shall encompass factors relating to the site and factors relating to the interaction between the site and the installation, for all operational states and accident conditions, including accidents that could warrant emergency response actions."

3.2.   Requirement 4 of SSR-1 [1] states that **"The suitability of the site shall be assessed at an early stage of the site evaluation and shall be confirmed for the lifetime of the planned nuclear installation."**

3.3.   Paragraph 4.6 of SSR-1 [1] states (citation omitted):

"In the assessment of the suitability of a site for a nuclear installation, the following aspects shall be addressed at an early stage of the site evaluation:

....

(b)   The characteristics of the site and its environment that could influence the transfer of radioactive material released from the nuclear installation to people and to the environment;

(c)   The population density, population distribution and other characteristics of the external zone, in so far as these could affect the feasibility of planning effective emergency response actions, and the need to evaluate the risk to individuals and to the population."

3.4.   Requirement 14 of SSR-1 [1] states:

"The data necessary to perform an assessment of natural and human induced external hazards and to assess both the impact of the environment on the safety of the nuclear installation and the impact of the nuclear installation on people and the environment shall be collected."

3.5.   Paragraph 4.46 of SSR-1 [1] states:

"At a minimum, the data collection process shall include the following:

....

(d)   Information on the potential impact of the nuclear installation on people and the environment for operational states and accident conditions;

(e)   Information required for planning effective emergency response actions on the site and off the site in all environmental conditions and for all states of the nuclear installation".

3.6.   To meet the requirements quoted in paras 3.2 and 3.3, the selected site for a new nuclear installation is expected to go through a characterization process. Depending on the complexity of the site, the data collection and investigations might take several years before the application for the construction licence for the proposed nuclear installation is submitted to the regulatory body. Data collection and investigations should continue during the construction, operation and decommissioning stages of the installation to confirm that the public and the environment continue to be protected and that the environmental impacts are in accordance with the predictions. This can be done as part of a review of the site evaluation within the framework of the periodic safety review. As stated in para. 4.48 of SSR-1 [1]:

"The data shall be maintained and reviewed periodically, and/or as necessary as part of a review of the site evaluation within the framework of the periodic safety review of the nuclear installation, for example, to address developments in data gathering techniques and in the analysis and use of data and to confirm that the data remain relevant to the site within the context of evolving hazards."

3.7. In order to assess the potential effects of the nuclear installation on the region, the site characterization efforts conducted prior to construction in compliance with Requirements 25–27 of SSR-1 [1] serve the following purposes:

(a) To establish the baseline environmental conditions at the selected site (using data quality objectives), which can later be used to measure the incremental environmental impacts of the nuclear installation during construction, operation and decommissioning;
(b) To apply calculational models for prospective radiological dose assessments;
(c) To evaluate the feasibility of planning effective emergency response actions;
(d) To establish the beginning of the monitoring programme at and in the vicinity of the site (see Section 8).

3.8. For the purpose of assessing the radiological environmental impact of a nuclear installation, background environmental data should be compiled for the following areas:

(a) Population distribution;
(b) Uses of land and water in the region of the site;
(c) Background radioactivity in environmental media;
(d) Meteorological characteristics of the region;
(e) Hydrological, geological and hydrogeological characteristics of the site catchment for surface water and groundwater.

The background environmental data needed for assessing non-radiological impacts (see para. 1.13), such as chemical, physical and socioeconomic impacts and the impact on culturally and historically significant properties at or near the site, are ideally compiled at the same time as the data needed for radiological impact assessments.

3.9.   Efforts should be made to collect data that allow transboundary impacts to be assessed. When a site is near a national border, there should be appropriate cooperation with the neighbouring countries in the vicinity of the nuclear installation. Efforts should be made to exchange relevant information.

3.10. The extent of the geographical area for which these data are compiled should be based on the anticipated effects of the environment on the safety of the nuclear installation intended to be built and the anticipated effects of the nuclear installation on the environment during operational states and in accident conditions. The geographical extent of the investigations should be wide enough to include both the peak radionuclide concentration and the maximum predicted dose plotted as a function of distance from the installation.

3.11. The spatial and temporal resolution of data collection activities should follow a graded approach, as described in Section 10. Consequently, more data should be collected for locations where installations with a higher radiological hazard are sited.

POPULATION DISTRIBUTION

3.12. Requirement 26 of SSR-1 [1] states:

"The existing and projected population distribution within the region over the lifetime of the nuclear installation shall be determined and the potential impact of radioactive releases on the public, in both operational states and accident conditions, shall be evaluated and periodically updated."

3.13. Paragraph 6.8 of SSR-1 [1] states:

"Information on the existing and projected population distribution in the region, including resident populations and (to the extent possible) transient populations, shall be collected and kept up to date over the lifetime of the nuclear installation. Special attention shall be paid to vulnerable populations and residential institutions (e.g. schools, hospitals, nursing homes and prisons) when evaluating the potential impact of radioactive releases and considering the feasibility of implementing protective actions."

3.14. Paragraph 6.9 of SSR-1 [1] states:

"The most recent census data for the region, or information obtained by extrapolation of the most recent data on resident populations and transient populations, shall be used in obtaining the population distribution. In the absence of reliable data, a special study shall be carried out."

3.15. Paragraph 6.10 of SSR-1 [1] states:

"The data shall be analysed to obtain the population distribution in terms of the direction and distance from the site. This information shall be used to carry out an evaluation of the potential radiological impact of normal discharges and accidental releases of radioactive material, including reasonable consideration of releases due to severe accidents, with the use of site specific design parameters and models as appropriate."

3.16. The distribution and characteristics of the regional population should be studied at the site characterization stage in order to evaluate the radiological impacts of discharges and accidental releases and to assist in evaluating the feasibility of planning effective emergency response actions.

3.17. Data on the present population[4] in the external zone[5] should be obtained from census data, from local authorities or by means of special field surveys, and those data should be as accurate and as up to date as possible. Similar data should also be collected for the region outside the external zone to distances determined in accordance with national practice and regulatory requirements and with the expected range of the impact of the project. The data should include the number of people normally present in the area, and the location of houses, hospitals, nursing homes, prisons, military bases, schools, industrial facilities, other institutions and recreational facilities such as parks and marinas.

3.18. The information collected on the present population should relate to the doses that might be received by individuals from direct radiation and from the release of radionuclides from the nuclear installation during operational states and in accident conditions. It should also address factors that would affect the

---

[4] 'Present population' includes both the resident population and the transient population.

[5] The 'external zone' is the area immediately surrounding a proposed site area in which population distribution and density, and land and water uses, are considered with respect to their impact on planning effective emergency response actions. This is the area that would be the emergency planning zone if the facility were in place [6].

feasibility of planning effective emergency response actions. This should include information on major places of work, means of communication, typical living habits (e.g. recreational and work activities, the fraction of time spent indoors versus outdoors) and typical diet of the inhabitants. Typical production rates of food items locally grown and the fractions locally consumed should be given.

3.19. The presence of a major industrial facility in a city or town in the region should be considered. For example, the facility might have a large workforce that would need to be evacuated in an emergency, or the facility might handle or process hazardous materials and need its own emergency arrangements that would need to be coordinated with the emergency plan of the nuclear installation.

3.20. The information collected on the transient population should cover the short term transient population (e.g. tourists, nomads) and the long term transient population (e.g. seasonal inhabitants, students). The maximum size of the transient population and its periods of occupancy in the external zone should be estimated. In the area outside the external zone, estimates of the approximate size of the transient population and its periods of occupancy should also be made.

3.21. A projection of the present population in the region should be made both for the expected year of commissioning of the nuclear installation and for selected years over the lifetime of the installation (e.g. every tenth year), if this is feasible and the population is expected to change appreciably in the future. Projections should be made on the basis of population growth rate, migration trends and plans for possible development in the region, including the project itself. The projected figures for resident population and transient population should be extrapolated separately if these population data are available.

3.22. The representative person associated with each nuclear installation should be identified (see paras 7.14–7.17 of this Safety Guide, section 5 of GSG-10 [13], and Ref. [19]).

3.23. The population data collected should be presented in a suitable format and at a suitable scale to permit their use with other relevant data, such as data on atmospheric dispersion and on uses of land and water. The data collected on the resident population and any transient populations should be in the form of numbers of people at specific locations represented as, for example, geographical coordinates, polar or Cartesian grid elements, or names of geographical features. In this way, the population data can be displayed on a map of the site and its surroundings at various scales; this can be done using a geographical information system (GIS). Collecting data at the highest spatial resolution possible allows greater flexibility in the future use of the data. The spatial resolution should be the highest, at least for areas closer to the site, especially within the external zone.

USES OF LAND AND WATER IN THE REGION OF THE SITE

3.24. Requirement 27 of SSR-1 [1] states that **"The uses of land and water shall be characterized in order to assess the potential effects of the nuclear installation on the region."**

3.25. The operation of a nuclear installation might affect the uses of land and water in the surrounding area. The availability of cooling water is an important consideration for the siting of a nuclear power plant. Therefore, as part of the site evaluation, the site topography (e.g. flat plains, mountains, hills, creeks, wetlands, valleys, forests) should be described and the uses of land and water should be investigated.

3.26. The investigations should cover the following aspects depending on their relevance for the site:

(a) Land devoted to agricultural uses, its extent and the main crops and their yields;
(b) Land devoted to dairy farming, its extent and its yields;
(c) Land devoted to industrial, institutional and recreational purposes, its extent and the characteristics of its use;
(d) Bodies of water used for commercial or recreational fishing, including details of the aquatic species fished, their abundance and their yields;
(e) Bodies of water used for commercial purposes (e.g. navigation), community water supply, irrigation and recreational purposes such as bathing and sailing;
(f) Land and bodies of water supporting wildlife and livestock;
(g) Direct and indirect pathways for potential radioactive contamination of the food chain;
(h) Products imported to or exported from the region that may form part of the food chain;
(i) Forest food and seaweed.

3.27. The characteristics of the land and water used in the region might affect the feasibility of planning effective emergency response actions (see Section 9).

3.28. Present uses of water that could be affected by changes in the water temperature and by radioactive substances discharged from a nuclear installation, together with the location, nature and extent of usage of the water, should be identified. Expected changes in uses of water in the region, such as for irrigation, fishing and recreational activities, should also be considered.

3.29. Special consideration should be given to any population centres for which drinking water is obtained from water bodies that might be affected by a nuclear installation. To the extent possible, future water flow and water uses should be projected over the lifetime of the installation. This might lead to changes in the representative person.

3.30. The movement and quality of groundwater should be studied to demonstrate the possibility and extent of groundwater contamination by radioactive material releases or leakages from the nuclear installation and to indicate whether groundwater could be a significant exposure pathway.

3.31. Different water uses should include the following data, depending on their relevance for the site:

(a)　For water used for drinking by humans and animals, and for municipal and industrial purposes:
　　(i)　Average and maximum rates of water intake by humans and animals;
　　(ii)　Distance of the intake from the potential source of radioactive discharges;
　　(iii)　Mode of water consumption;
　　(iv)　Number of water users.
(b)　For water used for irrigation:
　　(i)　Rate of water use;
　　(ii)　Area of irrigated land;
　　(iii)　Types and yields of agricultural products and their usual consumers.
(c)　For water used for fishing:
　　(i)　The aquatic species fished, and their abundance and yields in water used for commercial, individual and recreational fishing;
　　(ii)　The usual consumers of commercial fishing.
(d)　For water used for recreational purposes:
　　(i)　The number of persons engaging in swimming, sailing and other recreational uses;
　　(ii)　The time spent on these activities.

3.32. These investigations should cover a reasonably large area of the region where the site is located. The area should be identified according to the hydrological characteristics (relating to the basin or sub-basin) and hydrogeological characteristics (including possible recharge and discharge areas of the groundwater system) of the region where the site is located. The land area should encompass the region over which the peaks in pathway specific doses from various exposure routes are encountered. If the nuclear installation is located on a riverbank, users downstream from the site should be identified. If the site is near a lake, all users of the lake should be identified. If the site is on a coast, users of the sea out to a few tens of kilometres in all directions should be identified.

BACKGROUND RADIOACTIVITY IN ENVIRONMENTAL MEDIA

3.33. Paragraph 7.3 of SSR-1 [1] states:

"Before commissioning of the nuclear installation begins, the levels of background radioactivity in the atmosphere, hydrosphere and lithosphere and in biota in the region shall be measured so as to make it possible to determine any additional radioactivity due to the operation of the nuclear installation."

3.34. Any exposure to radiation from nuclear installations is additional to the exposure from the natural background. The calculation of the dose from a nuclear installation should not include the dose due to this natural background.

3.35. The importance of establishing the baseline radiation levels during the pre-operational stage of a nuclear installation is emphasized in IAEA Safety Standards Series No. GSG-9, Regulatory Control of Radioactive Discharges to the Environment [20]. A monitoring programme should be implemented to measure external radiation exposure (dose rates) and activity concentrations in environmental media. To assess the exposure due to the nuclear installation, the monitoring results due to the natural background should be subtracted.

3.36. The background radioactivity of a site area should be measured as early as possible, before the proposed nuclear installation starts operation and any release to the environment from the nuclear installation occurs. Information should be collected and recorded on the levels of background radioactivity and activity concentrations of relevant radionuclides in environmental media within the zone around the installation that is likely to be affected by any planned or potential releases from the installation, in particular in locations where exposure is expected to be higher (e.g. downwind from the proposed stack location, in sediments near outfalls from the proposed aquatic discharges). This zone should extend as far as necessary to include the location of the potential representative person; this distance depends on local site conditions. However, some of the environmental sampling locations should extend further to serve as control locations that could indicate potential changes in the composition of the background radioactivity during the operation of the installation.

3.37. If there are other sources of human-made radioactivity or enhanced natural radioactivity (e.g. another nearby nuclear installation or industrial facility, naturally occurring radioactive materials) that contribute to the radioactivity levels in the vicinity of the site, these should also be measured to determine the cumulative exposure of people around the site to human-made radiation. Any future discharges from these other sources should also be modelled, so that the total exposure of people in the area can be determined.

3.38. If there are any known existing exposure situations at the site or in the vicinity (e.g. from prior remediation activities or from a nuclear or radiological emergency that has been declared to be ended), those situations and the doses attributable to them should be identified. Requirements on existing exposure situations are established in GSR Part 3 [4].

3.39. The measurement programme set up to assess the background radioactivity in environmental media before the start of operation should guide the development of the long term monitoring programme to be followed during operation and decommissioning of the nuclear installation (see Section 8).

METEOROLOGICAL CHARACTERISTICS OF THE REGION AND DATA COLLECTION

3.40. Paragraph 6.2 of SSR-1 [1] states:

"A programme for meteorological measurements shall be prepared and carried out at or near the site using instrumentation capable of measuring and recording the main meteorological parameters at appropriate elevations, locations and sampling intervals. Data from at least one representative full year shall be collected and used in the analyses of atmospheric dispersion, together with any other relevant data available from other information sources. The meteorological data shall be expressed in terms of appropriate meteorological parameters."

3.41. The objective of meteorological investigations should be the continuous collection and evaluation of data on site specific meteorological parameters needed to calculate the atmospheric dispersion of radionuclides discharged from a nuclear installation during operational states or released in accident conditions (see Section 4). Another separate objective relating to meteorological data is the collection of data to derive extreme and rare meteorological hazards for the nuclear installation; this is addressed in SSG-18 [7] and not discussed further in this Safety Guide.

3.42. Investigations should be undertaken in the region of the site to collect specific meteorological information. This information should be compiled as part of the analysis and estimation of site specific values of meteorological parameters. Further recommendations are provided in SSG-18 [7].

3.43. Meteorological investigations should extend for at least one full year to account for seasonal variations, and for as long as needed to determine the representative meteorological conditions for the site (i.e. a year of data that have a similar pattern to that seen for the total period in terms of parameters such as wind speed and wind direction distribution, seasonal precipitation and atmospheric stability distribution). In order to have a full set of data for at least one representative full year, at least three years and ideally up to ten years of data should be collected. The collection of data should continue for the lifetime of the nuclear installation to confirm that conditions have not changed significantly and that updated assessments can be performed using the latest data as necessary. When a change is identified, a new assessment should be performed using the latest data.

3.44. The meteorological data collected should be compatible — in terms of their nature, scope and precision — with the methods and models in which they are used in evaluating the radiation exposure of the public and the radiological impact on the environment for assessment against each regulatory requirement (see Section 4).

3.45. Meteorological measurements are often affected by terrain and local features. The terrain in the range of several kilometres around a nuclear installation site should be examined, paying particular attention to topographical features (e.g. valleys, principal ridges, coastlines) and to plant structures (e.g. cooling towers, masts supporting meteorological sensors). Building wake effects might also influence the representativeness of the data obtained. In collecting meteorological data, care should be taken to prevent local effects from unduly altering the values of the parameters being measured.

3.46. For the purpose of atmospheric dispersion analysis, data on the following meteorological parameters should be obtained concurrently:

(a)   Wind vectors (i.e. wind direction and speed);
(b)   Precipitation;
(c)   Air temperature;
(d)   Humidity;
(e)   Air pressure;
(f)   Specific indicators of atmospheric turbulence.

3.47. The data collected should adequately represent local meteorological conditions, and the extent to which the data represent the long term meteorological characteristics of the site should be indicated. The local data collected should be compared with any available concurrent and long term data from synoptic meteorological stations in the surrounding area to determine long term trends for the site or, if the local results seem anomalous, to investigate possible causes.

3.48. Meteorological investigation activities should be undertaken in accordance with accepted international standards, such as those presented in Ref. [21]. If meteorological equipment is installed, it should be positioned so as to obtain data representing the dispersion conditions at projected or actual release points. Equipment should be unobstructed and should be positioned far enough from any obstacles to minimize their effects on measurements. Ground cover and vegetation should be managed for the duration of the investigation programme, so that they do not obstruct the operation of the equipment. The positions and settings of the equipment should be selected for maximum exposure to the meteorological conditions. Activities should be undertaken in accordance with accepted international standards, such as those presented in Ref. [22].

3.49. If the meteorological investigation is to be conducted for a new installation at an existing site and there is a significant distance between the meteorological equipment of the existing facility and the proposed location of the new installation, it might be appropriate to conduct a validation study to utilize the existing meteorological data. This validation should be based on measurements made at the location of the new installation at a scale that provides a high level of confidence.

3.50. To conduct the programme for meteorological measurements, a meteorological tower should be installed at the site, or access to necessary data from a nearby, reliable source should be available. To ensure the measuring of meteorological parameters at appropriate elevations and to obtain realistic dispersion parameters, data should be collected at least 10 m above ground (to allow for comparison with the data from the synoptic meteorological stations) and at the height of any proposed stack (to be determined on the basis of preliminary information).

3.51. Meteorological measurements should be made at more than one location. For example, where the effect of sea breezes is important, data from an additional meteorological station further inland should be used in order to evaluate characteristics of the diffusion regime for the sea breeze over land.

3.52. Meteorological data should be obtained at least hourly. The averaging time and the sampling time for the data should be in accordance with the regulatory requirements. The instruments should provide continuous recording so that the data collected are readily available in situ to provide the hourly data. The raw data should be stored until data qualification and statistical analysis have been performed. Hourly mean values derived from the programme for meteorological investigation should be stored for the lifetime of the installation. Data averaged over shorter periods of time (less than one hour) should be stored continuously for purposes of emergency response and recovery, as they can be used to assess the plume dispersion in the event of an accidental release.

3.53. Fluctuations in meteorological conditions are direct indicators of atmospheric turbulence. To support the computational models used for atmospheric dispersion in determining the atmospheric stability, data relating to the following parameters should be obtained (see also Section 4):

(a)   Direct turbulence measurements (e.g. using a Doppler radar);
(b)   Fluctuations in wind direction;
(c)   Air temperature and temperature lapse rate;
(d)   Wind speed combined with solar radiation levels or sky cover during the daytime plus sky cover or net radiation levels during the night time.

3.54. For some computational models, the height of a mixing or boundary layer may need to be determined (see also paras 4.35 and 4.39).

3.55. For regions with complex meteorological conditions, for example in relation to mountainous regions, measurements of turbulence indicators made at the site alone might not be sufficient. Depending on the particular characteristics of the region, it may be necessary to take additional measurements of wind and turbulence indicators a few kilometres from the site. Normal discharges of effluents or experimental discharges of tracers can be used for the development of a local diffusion model, which is often a general model with adjustments derived from air concentration values measured at the site and in the region.

3.56. In developing site specific dispersion models, sufficient information should be acquired on the spatial and temporal distributions of parameters relating to wind direction and speed and atmospheric stability, to be able to understand and determine the trajectory of effluents. Such information could be obtained by way of a programme of field measurements and/or by the method of numerical weather prediction (see para. 3.59).

3.57. Precipitation should be reported at least hourly. Measurements of the intensity of precipitation and total precipitation as well as details of the type of precipitation should be used to evaluate the impact of precipitation on airborne concentrations of contaminants and on ground contamination. Data on humidity may also help to determine any effects of cooling towers (e.g. icing or fogging on roadways and bridges, visibility of cooling tower plumes, effects of salt drift on vegetation). Air humidity can modify the dispersion of aerosols, as it can increase the coalescence of particulates.

3.58. The wind speed and direction at different elevations and temperatures should be averaged at least once per hour, while for other variables such as solar radiation levels and precipitation levels, the period of integration should be one hour. Wind direction should be averaged as a vector (e.g. the use of circular mean from circular statistics), as well as a scalar value, to support multiple potential modelling methods.

3.59. Local meteorological data can also be derived from dynamic numerical atmospheric prediction models. In this method, data collected (usually by the national meteorological institute) from many sources (e.g. a network of land stations, satellites and observations from commercial aircraft, ships, buoys, radiosondes) are processed by the same models used for numerical weather prediction and interpolated for a specified location. In this way, hourly meteorological data can be obtained retrospectively for any location within the region over which the data have been collected and for as far back historically as the data exist.

3.60. The programme for regional meteorological investigation and all information relating to it should be documented for the purposes of site evaluation and design and for use in emergency plans.

HYDROLOGICAL, GEOLOGICAL AND HYDROGEOLOGICAL CHARACTERISTICS OF THE SITE CATCHMENT FOR SURFACE WATER AND GROUNDWATER

3.61. Paragraph 6.7 of SSR-1 [1] states:

"The hydrogeological and hydrological investigations shall determine, to the extent necessary, the dilution and dispersion characteristics of water bodies, the re-concentration ability of sediments and biota, the migration and retention characteristics of radionuclides, the transfer mechanisms for radionuclides in the hydrosphere, as well as the associated exposure pathways."

3.62. As part of establishing the environmental background and baseline conditions at a site, the hydrological and hydrogeological characteristics of the region around the proposed nuclear installation should be investigated. Emphasis should be placed on investigations relating to the surface water and groundwater flows within the catchment area where the nuclear installation is located, keeping in mind the potential effects of water flow on the proposed installation and the potential effects of the proposed installation on the quantity and quality of water in the region. The region for investigations should be extended beyond the catchment area if the discharges to the atmosphere could reach and deposit in significant quantities on the ground and on surface water there.

3.63. The geology and surface hydrology of the site area should be studied in sufficient detail (including the factors that might affect surface hydrology, such as changes in land use and meteorological conditions) to indicate potential pathways for the transfer of radionuclides to surface water or groundwater. Any surface drainage system or standing water body accessible from a potential release point in an accident should be identified. Areas from which contaminated surface water might directly enter an aquifer should be determined. The relevant hydrogeological information for surface or near surface discharges includes information on soil moisture properties, infiltration rates, configuration of unsaturated zones and chemical retention properties under unsaturated conditions, as necessary.

Surface water

3.64. Typical surface water bodies in the vicinity of a nuclear installation range from rivers to inland freshwater lakes (natural or human-made) to marine systems (e.g. estuaries, seas, oceans). The distinction between natural and human-made lakes needs to be made since the natural dynamics of human-made lakes are altered by operational use (e.g. hydropower generation, irrigation). Recommendations on the collection of background hydrological data for sites on these types of water bodies are provided in paras 3.66–3.71.

3.65. Recommendations on the parameters needed to analyse the dispersion of radionuclides in surface water are provided in Section 5.

Rivers

3.66. For sites on rivers, the collection of hydrological and other information should cover the following:

(a) The channel geometry defined by the mean width, the mean cross-sectional area and the mean slope over the river reaches of interest. (The water level can be computed from the channel geometry and the river flow rate.) If there are important irregularities such as dead zones or hydraulic equipment in the stream that could influence the dispersion of the plume, these should be described. Additional downstream measurements of channel geometry should be made as necessary to assess the dispersion process in the river.

(b) The river flow rate presented as monthly averages of the inverse of daily flows. The inverse rate of flow should be used, since the fully mixed concentration is proportional to the reciprocal of the flow rate if sediment sorption effects are not considered. The flow rates of other relevant and important water bodies (e.g. downstream tributaries of the river) should be measured if they affect dispersion. Although monthly flow averages are recommended for generic analysis, daily and even instantaneous stream flow data might have, if any, some effects on estimated dose assessments. The use of instantaneous flow data might result in higher flow rates for the river system in question.

(c) Extremes in the flow rate evaluated from available historical data.

(d) Seasonal variation of the water level over the reaches of interest.

(e) Tidal variations in water level and flow rate in the case of a tidal river.

(f) Data describing possible interactions between river water and groundwater, and the identification of those reaches of the channel where the river might gain water from or lose water to groundwater.

(g) River temperature, measured at representative locations (e.g. one location representative of upstream and one representative of downstream in the river) over at least a year and expressed as monthly averages of daily temperatures.

(h) The thickness of the top water layer, if thermal stratification of water in the river occurs.

(i) Extreme temperatures evaluated from available historical data.

(j)   The concentrations of suspended matter measured:
      (i)   At locations downstream of sections where the river is slowed,
            depleted or fed by tributaries;
      (ii) In discrete samples at appropriate time intervals (e.g. every two
            months for at least a year);
      (iii) Over a sufficient range of flows to establish curves of flow versus
            sedimentation and/or erosion rate.
(k)   The characteristics of deposited sediments, including mineral and/or organic
      compositions and size classification.
(l)   The distribution coefficients for sediments and for suspended matter for the
      various radionuclides that might be discharged.
(m)   The background levels of radioactivity in water, sediment and aquatic food
      due to natural and artificial sources.
(n)   Seasonal cycles of phytoplankton and zooplankton, with at least the periods
      of their presence and cyclical evolutions of their biomass.
(o)   Spawning periods and feeding cycles of major fish species.
(p)   Dams located upstream and the water volumes in the adjacent lakes.

Natural lakes

3.67. The natural lakes that are used as a source of cooling water for nuclear
power plants tend to be large lakes. The information to be collected for such lakes
should include the following:

(a)   The general shore and bottom configuration in the region, and unique
      features of the shoreline in the vicinity of the discharge.
(b)   Data on bathymetry out to a distance of several kilometres, and data on the
      amount and character of sediments in the shallow shelf waters.
(c)   Speeds, temperatures and directions of any near shore currents that could
      affect the dispersion of discharged radioactive material. Measurements
      should be made at appropriate depths and distances, depending on the
      bottom profile and the location of the point of discharge.
(d)   The duration of stagnation and characteristics of current reversals. After
      stagnation, a reversal in current usually leads to a large scale mass exchange
      between inshore and offshore waters that effectively removes pollutants
      from the shore zone.
(e)   The thermal stratification of water layers and its variation with time,
      including the position of the thermocline and its seasonal changes.
(f)   The load of suspended matter, sedimentation rates and sediment distribution
      coefficients, including data on sediment movements characterized by
      defining at least the areas of high rates of sediment accumulation.

(g)   The background levels of radioactivity in water, sediment and aquatic food due to natural and artificial sources.

(h)   Seasonal cycles of phytoplankton and zooplankton, with at least the periods of their presence and cyclical evolutions of their biomass.

(i)   Spawning periods and feeding cycles of major fish species.

(j)   If applicable, the extent of seasonal ice formation (thermal plumes can affect ice cover and fish spawning areas).

Human-made lakes

3.68. For sites on human-made lakes, the information to be collected should include the following:

(a)   Parameters of the lake geometry, including length, width and depth at different locations;

(b)   Rates of inflow and outflow (considering the operational programme);

(c)   Expected fluctuations in water level on a monthly basis;

(d)   The water quality at inflows, including temperature and suspended solids;

(e)   The thermal stratification and its seasonal variations;

(f)   Interaction with groundwater;

(g)   Characteristics of bottom sediments (e.g. type and quantity);

(h)   The distribution coefficients for sediments and for suspended matter for the various radionuclides that might be discharged;

(i)   The rate of sediment deposition;

(j)   The background levels of radioactivity in water, sediment and aquatic food due to natural and artificial sources;

(k)   Seasonal cycles of phytoplankton and zooplankton, with at least the periods of their presence and cyclical evolutions of their biomass;

(l)   Spawning periods and feeding cycles of major fish species;

(m)   If applicable, the extent of seasonal ice formation (thermal plumes can affect ice cover and fish spawning areas).

Estuaries

3.69. For sites on estuaries, the following information should be collected:

(a)   The salinity distribution determined along several verticals covering different cross-sections of the salinity intrusion zone. The data should be sufficient to delineate the flow pattern, which is directed towards the estuary mouth in the upper layer and towards the inner reaches in the lower layer of a fully or partially mixed estuary.

(b)   Evaluations of sediment displacements, the load of suspended matter, the rate of buildup of deposited sediment layers and the movement of these sediments with the tide.

(c)   Channel characteristics sufficiently upstream of the site to model the maximum upstream travel of radioactive effluents, if applicable.

(d)   The distribution coefficients for sediments and for suspended matter for the various radionuclides that might be discharged.

(e)   The background levels of radioactivity in water, sediment and aquatic food due to natural and artificial sources.

(f)   Seasonal cycles of phytoplankton and zooplankton, with at least the periods of their presence and cyclical evolutions of their biomass.

(g)   Spawning periods and feeding cycles of major fish species.

3.70. Measurements of water temperature, salinity and other relevant water quality parameters in estuaries should be made at appropriate depths, distances and times, depending on the river flow, tidal levels and the configuration of the water body in different seasons.

Open shores of seas and oceans

3.71. For sites located on the shores of seas and oceans, the information to be collected should include the following:

(a)   The general shore and bottom configuration in the region, and unique features of the shoreline in the vicinity of the discharge.

(b)   Data on bathymetry out to a distance of several kilometres, and data on the amount and character of sediments in the shallow shelf waters.

(c)   Speeds, temperatures and directions of any near shore currents and tides that could affect the dispersion of discharged radioactive material. Measurements should be made at appropriate depths and distances, depending on the bottom profile and the location of the point of discharge.

(d)   The duration of stagnation and characteristics of current reversals. After stagnation, a reversal in current usually leads to a large scale mass exchange between inshore and offshore waters that effectively removes pollutants from the shore zone.

(e)   The thermal stratification, if it exists within a reasonable distance from the shoreline, of water layers and its variation with time, including the position of the thermocline and its seasonal changes.

(f)   The load of suspended matter, sedimentation rates and sediment distribution coefficients, including data on sediment movements characterized by defining at least the areas of high rates of sediment accumulation.

(g)   The background levels of radioactivity in water, sediment and aquatic food due to natural and artificial sources.

(h)   Seasonal cycles of phytoplankton and zooplankton, with at least the periods of their presence and cyclical evolutions of their biomass.

(i)   Spawning periods and feeding cycles of major fish species.

Groundwater

3.72. A conceptual model of the hydrogeological conditions at and around the site where the installation is proposed should be developed. To understand the transport and fate of contaminants, the hydrogeological conceptualization and characterization should include a description of the groundwater system in terms of the type of flow domain (e.g. intergranular, fractured, karstic), flow regime (e.g. confined, unconfined, leaky) and hydrological zones (e.g. vadose zone, phreatic zone). This conceptual model should indicate the following (see also Section 6):

(a)   Hydrostratigraphical description of lithological units;
(b)   Water inflows and outflows;
(c)   Connectivity and interaction between the surface water bodies and groundwater;
(d)   Spatial distribution of potentiometric level and groundwater flow direction.

3.73. The information to be collected, on a local and regional scale, to identify the hydrogeological system should include:

(a)   Climatological and hydrological data;
(b)   Initial concentrations of radionuclides;
(c)   Major hydrogeological units, their hydraulic parameters and the ages or mean turnover times of groundwater;
(d)   Recharge and discharge relationships.

3.74. In terms of climatological data, in regions where rainfall makes a substantial contribution to groundwater, hydrometeorological data on daily and monthly rainfall and the data needed to calculate the potential and actual evapotranspiration that have been systematically collected should be analysed throughout the period for which they are available. From the precipitation data, groundwater recharge should be calculated. Alternative methods such as tracers (chemical or isotopic) of the water cycle could be introduced to calculate groundwater recharge.

3.75. Data should be obtained on the various types of geological formations in the region and their stratigraphical distribution in order to characterize the regional groundwater system and its relationship with the local hydrogeological units. These data should include the following:

(a) Geological data: lithology, thickness, faults and fracture systems, extent and degree of homogeneity of the geological units;
(b) Hydrogeological data: description of the unsaturated zone, hydraulic conductivities and transmissivities, specific yield and storage coefficients, dispersivity, hydraulic gradients of the saturated zone for the geological units that form a flow domain, and an inventory of wells used around the site with their chronicles and pumping rates;
(c) The chemical composition of groundwater from the respective aquifers;
(d) Variations of water levels in wells and in the discharges of springs and rivers;
(e) Morphological features in karstic terrains, locations of closed depressions and active and potential sinkholes in the region.

3.76. For the relevant hydrogeological units, information should be collected on the following chemical and physical properties of the groundwater:

(a) Physical properties of groundwater (e.g. temperature, electrical conductivity, turbidity);
(b) Chemical properties of groundwater (e.g. pH, redox potential);
(c) Concentrations of major anions and cations;
(d) Sorption characteristics, when necessary for the selected grade of modelling.

4. ANALYSIS OF THE DISPERSION OF RADIONUCLIDES IN THE ATMOSPHERE

4.1. Paragraph 6.1 of SSR-1 [1] states:

"The analysis of the atmospheric dispersion of radioactive material shall take into account the orography, land cover and meteorological features of the region, including parameters such as wind speed and direction, air temperature, precipitation, humidity, atmospheric stability parameters, prolonged atmospheric inversions and any other parameters required for modelling of atmospheric dispersion. If possible, long term meteorological data for nearby locations shall be obtained, evaluated for quality and used."

4.2. Dispersion of released radionuclides in the atmosphere is a major exposure pathway by which radioactivity that is either routinely discharged under authorization or accidentally released from a nuclear installation could be transported to locations where the public could be exposed. Exposure can be immediate, by inhalation of material in the plume or by external exposure from radiation in the cloud, or it can occur over an extended period of time from material deposited on the ground, which can cause external exposure, or from ingestion of radioactivity incorporated in the food chain.

4.3. Releases of radioactivity from nuclear installations to the atmosphere — or to the environment in general — occur as part of planned operation. These can be authorized or permitted routine discharges, including releases from anticipated operational occurrences (a deviation of an operational process from normal operation that is expected to occur at least once during the operating lifetime of a facility), or releases resulting from events that are not expected to occur with certainty, such as accidents. Such releases correspond to a planned exposure situation, including potential exposure.

4.4. For modelling the dispersion of radioactivity in the environment, a distinction should be made between continuous releases and short term releases, since each type of release involves a different approach. For example, routine or normal discharges might not be constant but might vary during the operational cycle or there might be spikes in the release during shutdown or maintenance, while some anticipated operational occurrences might not result in any release at all.

4.5. All releases should be subject to regulatory control, with normal operation (including anticipated operational occurrences) and accidental releases considered separately (see Section 7). Although routine releases are unlikely to be constant and might include short term and long term variability, for the purposes of applying for a permit or authorization, a constant continuous release is usually assumed based on the predicted releases plus a margin to allow for additional operational freedom. Some States might also consider some short term releases separately as part of normal operation.

4.6. Generally, different dispersion models are used to assess the potential impacts from continuous discharges and short term releases. However, the meteorological data that need to be acquired for each type of model are usually the same. In both cases, typical rather than extreme data are used, which can be collected from the site itself or by using numerical weather prediction models if sufficient quality data are available. However, as accidental releases are typically short term and might be potentially significant, assessments of accidental releases should consider the likelihood and potential effects of unusual meteorological conditions that could lead to higher doses. As mentioned in para. 3.43, the meteorological site data should be collected over several years so that it is possible to select a representative year or years from the records. The extreme data used for external hazard analysis need to come from a data set collected over a much longer period and typically only available on a regional basis. For long range dispersion analysis (which is typically needed for evaluating societal impacts or transboundary impacts), time and spatially gridded data for use in Lagrangian modelling (see para. 4.26) might need to be acquired from national or international meteorological organizations.

4.7. Short term releases can occur at any time under any meteorological conditions, and this can result in quite different radiological consequences depending on the conditions (e.g. as a result of different wind directions). One way to assess the consequences is to perform multiple calculations for different meteorological conditions sampled from the meteorological data set collected hourly and to apply statistical methods to the results.

4.8. The calculations of the dispersion and concentration of radionuclides coupled with the calculations of dose should show whether the radiological consequences of routine discharges and potential accidental releases of radioactive material into the atmosphere are tolerable. The results of these calculations for normal operation, including anticipated operational occurrences, should be used to establish authorized or permitted limits for radioactive discharges from the installation into the atmosphere (see GSG-9 [20]) and those from accident conditions, along with their expected frequency and comparison with national requirements for risk and/or dose criteria. In both cases, the results may be used to inform the design process to optimize doses or mitigate risks.

4.9. The results of the meteorological investigation should be used for the following purposes:

(a) To confirm the suitability of a site;
(b) To provide the baseline environmental data for site evaluation;
(c) To determine whether local meteorological characteristics have altered since the site evaluation was made and before operation of the installation commences;
(d) To select appropriate dispersion models for the site;
(e) To establish limits for radioactive discharges into the atmosphere;
(f) To establish limits for design performance (e.g. containment leak rates, control room habitability);
(g) To assist in evaluating the feasibility of planning effective emergency response actions.

SELECTION OF RELEASE SCENARIOS FOR DISCHARGES AND ACCIDENTAL RELEASES TO THE ATMOSPHERE

4.10. The release scenario for normal operation is the discharge — usually assumed to be a constant continuous discharge — from a known discharge location. Other scenarios to consider might include maintenance or other events, including anticipated operational occurrences, expected to occur during operation and that might lead to a short term release and a different profile of release in terms of radionuclides and/or their quantities.

4.11. A range of possible accident scenarios should be analysed (depending on the complexity of the installation), spanning from high frequency, low consequence events to low frequency, high consequence events.

4.12. For a nuclear power plant, this analysis might encompass many scenarios to cover the spectrum of events that might occur — anticipated operational occurrences, design basis accidents and design extension conditions, including severe accidents. For some advanced reactors, the definition of certain plant states might need to be adapted. Releases resulting from severe accidents might encompass the performance of a full scope Level 2 probabilistic safety assessment[6] (see SSG-4 (Rev. 1) [18] and IAEA Safety Standards Series No. SSG-3 (Rev. 1), Development and Application of Level 1 Probabilistic Safety Assessment for Nuclear Power Plants [23]). If a Level 2 assessment is incomplete during the site evaluation process, a release scenario may be based on a deterministic approach, accounting for proper uncertainties. The deterministic analysis should be verified once the Level 2 assessment has been completed.

4.13. For sites with multiple units and installations with multiple facilities, the site should be evaluated for interactions between the nuclear installations. There might be multiple discharges from several locations that all need to be analysed. Accident conditions might also include scenarios involving releases from multiple

---

[6] Three levels of probabilistic safety assessment are generally recognized:
  — Level 1 comprises the assessment of failures leading to determination of the frequency of fuel damage.
  — Level 2 includes the assessment of containment response, leading, together with Level 1 results, to the determination of frequencies of failure of the containment and release to the environment of a given percentage of the reactor core's inventory of radionuclides.
  — Level 3 includes the assessment of off-site consequences, leading, together with the results of Level 2 assessment, to estimates of public risks [6].
  See paras 1.4 and 1.5 of SSG-3 (Rev. 1) [23] for more details.

units that are either simultaneous or offset in time, which again might need to be analysed if these releases are significant contributors to the overall risk.

4.14. For nuclear installations with low potential hazard, the analysis of a few or even only one potential exposure pathway might be sufficient.

SELECTION OF SOURCE PARAMETERS FOR ANALYSING THE DISPERSION OF RADIONUCLIDES IN THE ATMOSPHERE

4.15. For discharges, the calculation of the source terms may necessitate a detailed analysis of fuel performance and the chemical regime for water cooled reactors, for example the generation of corrosion and activation products and the performance of waste treatment measures and filter systems.

4.16. Source terms for accident conditions differ from those for normal operation and anticipated operational occurrences (see SSG-2 (Rev. 1) [16] and IAEA Safety Standards Series No. SSG-90, Radiation Protection Aspects of Design for Nuclear Power Plants [24]). For example, they might involve larger quantities, different radionuclides and/or different physical and chemical characteristics.

4.17. Therefore, when selecting the source term parameters, consideration should be given to the following:

(a) Physical and chemical processes occurring during the accident sequence;
(b) Behaviour of any safety features or the effects of any mitigatory measures;
(c) Behaviour of radionuclides within the installation before they are released to the environment.

4.18. In addition to the quantities of radionuclides released, all the parameters that might affect the subsequent dispersion or behaviour of the radionuclides in the environment should be characterized. This characterization should include the following:

(a) Physical form (e.g. gas, aerosol);
(b) Chemical form;
(c) Release point and its height;
(d) Momentum and thermal energy associated with the release necessary to determine the effective height of the radioactive plume;
(e) Time profile for the release.

4.19. Sources of radioactivity in a nuclear power plant might include the following:

(a) Activated corrosion products that remain in the coolant during normal operation but that can be released to the environment in a loss of coolant accident (e.g. ^{58}Co, ^{60}Co).
(b) Fission products and actinides formed by fission or activation of uranium in fuel (e.g. ^{131}I, ^{137}Cs, ^{90}Sr, ^{238}Pu, ^{239}Np), and noble gases (^{85}Kr, ^{138}Xe). These are prevented from release in normal operation by many barriers.
(c) Radionuclides from the fuel matrix, fuel cladding, coolant circuit or containment. Volatile radionuclides can be released into the coolant through fuel rod failures or by a uranium contamination remaining on the outside surface of the cladding from the manufacturing process and/or from uranium impurities in the zirconium used in cladding in some fuel types ('tramp uranium'). The radionuclides can therefore be released when coolant is released or by off-gassing during normal operation. Large releases can also occur in severe accidents, when the fuel matrix and fuel cladding fail and the coolant circuit and containment might be breached.
(d) Activation products formed by the activation of substances present in the coolant water, which can be released when coolant is released or by off-gassing (e.g. ^{3}H, ^{14}C).

4.20. Radionuclides can also be released through fuel handling faults, radioactive waste handling faults or accidents involving waste or effluent storage.

4.21. In order to determine the source term, the fraction of the initial inventory released to the environment needs to be evaluated; if there are several barriers to the release, then the fraction released through each barrier needs to be assessed or modelled, as well as the processes that might lead to the mobilization of the source term. For potential exposures from nuclear power plants, numerical codes should be used (see also SSG-2 (Rev. 1) [16] and SSG-4 (Rev. 1) [18]).

4.22. Source term data should be supported by well documented numerical modelling and physical assumptions.

ATMOSPHERIC DISPERSION MODELS

4.23. Different tools are usually used for assessing the impacts of continuous and short term releases, although they may be based on the same type of atmospheric dispersion model. For continuous releases, the activities in the environmental compartments modelled increase until they reach an equilibrium level unless the radionuclide half-lives are long compared with the operational lifetime of the facility. For short term releases, the activity levels in the environmental compartments rise to a peak and then decay.

4.24. The most important factor in atmospheric dispersion modelling is to ensure that the model can simulate the important processes with sufficient accuracy. The atmospheric dispersion computer models commonly used fall into two main types: Gaussian and Lagrangian.

4.25. The Gaussian model describes the atmospheric dispersion by the Gaussian equation in both the crosswind and vertical dimensions. The plume spread increases with distance from the release according to the meteorological parameters such as wind speed and the parameters for atmospheric stability. The simple equations can be modified to take account of surface roughness, wet and dry deposition, building wake effects and plume rise.

4.26. A Lagrangian type model tracks many notional particles (i.e. hypothetical packets of air), transferring each by the wind direction and wind speed and then modelling atmospheric turbulence effects by applying random motion. The atmospheric dispersion parameters used to determine the motion of each particle are interpolated in space and in time from a three dimensional spatial grid of time series data. The concentration in any grid element of the model is determined by the number of particles in the box representing that grid element. Utilizing these gridded data allows modelling to vary meteorological conditions by location, height and time. Effects due to surface topography are implicitly considered as they are embedded in the meteorological data, provided that the data have sufficient spatial resolution.

4.27. The advantages of the Gaussian model are as follows:

(a)   It is a simple mathematical expression that is easy to implement.
(b)   It can be modified to take into account, in a simple way, effects such as plume rise, building wake effects and wet and dry deposition.
(c)   It is fast to execute so there is no need to sample from a meteorological data set; it is quite feasible to perform calculations for every hour of a data set of several years.
(d)   It is principally considered to be conservative compared with more detailed models. However, under specific conditions (e.g. for locations close to the location of the release), the results might not be conservative.

(e)   It needs a relatively simple meteorological data set of hourly data for the point of release, comprising data such as wind direction, wind speed, atmospheric stability category, mixing layer height and precipitation.

(f)   It can be adapted to model temporal and spatial changes in meteorological conditions during the release (Gaussian puff models).

(g)   More advanced Gaussian models have the ability to take account of more complex terrain and buildings in the vicinity of the release. A representative array of permanent surface meteorological stations should be considered for such models.

4.28. The disadvantages of the Gaussian model are as follows:

(a)   Other than the more advanced Gaussian models mentioned in para. 4.27(g), it cannot satisfactorily model complex terrain.

(b)   The range of validity is limited to the area over which the meteorological conditions remain reasonably constant and consequently it cannot satisfactorily model long range impacts.

(c)   Calm or very low wind speed conditions can only be modelled with high uncertainties.

(d)   The rate of dry deposition of material is dependent on particle size assumptions and might not satisfactorily model all release conditions.

4.29. The advantages of the Lagrangian model are as follows:

(a)   It can model complex terrain and long range dispersion if an appropriate meteorological data set is available (i.e. fine enough spatial resolution for the terrain being modelled to be resolved in the meteorological data) and therefore generally gives a more accurate representation of the dispersion for a given set of meteorological conditions.

(b)   It can model changes in meteorological conditions over time and space.

4.30. The disadvantages of the Lagrangian model are as follows:

(a)   It has long computer run times.

(b)   It needs a large complex meteorological data set including three dimensional data as a function of time for the area over which the model is being run.

(c)   Compromises may be needed in other areas to reduce the overall computation effect, such as the number of radionuclides or meteorological sequences modelled and geographical resolution.

(d)   It is difficult to use for continuous releases over an extended period.

4.31. A Lagrangian model could use single site data rather than a three dimensional grid but in this case — other than being less conservative — it offers few benefits over a simple Gaussian model. To take full advantage of the capabilities of a Lagrangian model, an extensive data set is needed, which should be collected from the national meteorological institute or from international sources of global data.

4.32. The decision on which model to use depends on the type of analysis needed and the characteristics of the site and surrounding area. If only an assessment of short range impacts is needed, and the surrounding area is reasonably flat, then the Gaussian model approach may be sufficient. If the surrounding area has complex topography, if long range results are needed for a transboundary assessment, or if an assessment of population risks in a large area is needed, the Lagrangian model may be more appropriate, assuming the necessary meteorological data of an appropriate resolution are available. For sites with complex topography and short range analysis, the more advanced Gaussian model could be used. For modelling dispersion in urban environments, more detailed approaches, such as those considering building heights and employing computational fluid dynamics, may be needed. There might still be large uncertainties associated with the source term, especially for accidental releases. Consequently, the additional insights gained from performing more sophisticated or more extensive analysis might not be commensurate with the additional effort and the decision on which model to use should be carefully evaluated.

USE OF METEOROLOGICAL AND OTHER DATA COLLECTED FOR MODELLING ATMOSPHERIC DISPERSION

4.33. Regardless of the means by which they have been acquired, the data should be compatible (in terms of their nature, scope and precision) with the methods and atmospheric dispersion models being used (i.e. Gaussian or Lagrangian models; see para. 4.24). For example, atmospheric stability can be characterized in different ways with different parameters. The data and models needed also depend on the regulatory requirements for the radiological impact on people or on the environment. For example, if an objective is to assess the risk to a whole population, then long range dispersion modelling is needed.

4.34. Generally, the same data collected for continuous releases can be used for short term releases. However, short term releases may also necessitate more long range data if long range dispersion modelling is part of the assessment — as might be the case for releases large enough to have an impact at long range.

4.35. The typical meteorological data needed for a Gaussian dispersion model include the following:

(a) Wind speed;
(b) Wind direction;
(c) Boundary (or mixing) layer height;
(d) Parameter(s) determining the atmospheric stability, such as the Pasquill–Gifford stability class, Doury scheme or Monin–Obukhov similarity theory;
(e) Precipitation.

4.36. Other data that may be used in the Gaussian dispersion model include the following:

(a) Deposition velocities for aerosol particles or chemical species to model dry deposition;
(b) Washout coefficients for aerosol particles or chemical species to model wet deposition;
(c) Release height;
(d) Energy or momentum of the release;
(e) Building dimensions to account for building wake effects;
(f) Surface roughness;
(g) Surface topography for more advanced Gaussian models.

Wind speed

4.37. The wind speed and atmospheric stability are related (see para. 4.40). Higher wind speeds might also have the effect of inhibiting plume rise effects. Higher wind speeds mean that radionuclides reach locations quicker, providing less time for any radioactive decay, but this is not usually significant unless very short lived radionuclides (e.g. with a half-life comparable to the time taken for the radioactivity to reach a particular location) are involved.

Wind direction

4.38. Wind direction can be very important if there is an uneven population distribution around the site, since the probability of exposure at any given location depends on the probability that the wind blows in that direction.

Boundary layer height

4.39. The boundary (or mixing) layer height is the height at which a temperature inversion occurs, creating an effective boundary for dispersion in the vertical direction. Gaussian dispersion models generally assume that the plume reflects down from the boundary layer with no transfer across the boundary, and up from the ground until fully mixed in the vertical direction. Lagrangian dispersion models may model transfer across the boundary and subsequent dispersion of material above the layer. The boundary layer height is important since it effectively determines the volume of air into which the plume of radioactivity can disperse as it moves downwind to a point of being fully mixed vertically. At these distances, the boundary layer height is correlated with the atmospheric stability, with more unstable conditions leading to higher boundary layer heights.

Atmospheric stability

4.40. Atmospheric stability is usually the most important atmospheric dispersion parameter after wind direction and should be considered in modelling. Unstable conditions lead to increased dispersion in both the vertical and crosswind directions and hence lower ground-level concentrations. For elevated releases (e.g. from a stack), unstable conditions increase the vertical dispersion, which causes the plume to reach ground level at shorter downwind distances than would be the case for more stable conditions, and can lead to higher concentration close to the release. For ground-level releases, more stable conditions lead to higher plume centreline concentrations but with lower crosswind spread.

Precipitation

4.41. Precipitation enhances the deposition of activity on the ground by washing material out of the plume. Precipitation can also transfer activity to surface water and/or groundwater systems and should be carefully modelled.

SENSITIVITY STUDY OF THE ANALYSIS OF RADIONUCLIDE DISPERSION IN THE ATMOSPHERE

4.42. Since an assessment involves many assumptions and uncertainties, sensitivity studies should be performed to assess the sensitivity of the overall results to the assumptions or parameter values. Typical sensitivity studies should include the following:

(a)   Meteorological data used (for practical reasons, only a subset or sample of the meteorological data may have been used, and the effect of different or larger samples could be investigated);

(b)   Source term assumptions, including activity released, possible release heights, energy of release and time profile of release;

(c)   Modelling uncertainties;

(d)   Representative person assumptions (e.g. age group, location, food consumption);

(e)   Parameter value assumptions (e.g. deposition velocities);

(f)   Assumptions about protective actions applied.

SCENARIO BASED SIMULATION OF ATMOSPHERIC DISPERSION

4.43. The scenarios simulated for routine or normal discharges should normally assume that the release continues at a constant rate for the lifetime of the nuclear installation, using hourly meteorological data. In addition, short term planned discharges, such as those that occur during maintenance, should also be modelled.

4.44. For radioactive releases from an accident at a nuclear installation, the site specific variability in the meteorological data and exposure scenarios may be accounted for in the deterministic assessment and, if needed, in the probabilistic assessments depending on the off-site consequence criteria to be met (see paras 7.18–7.20).

4.45. Such analyses may also be used to inform emergency planning or for transboundary assessments.

GRADED APPROACH TO ASSESSING THE DISPERSION OF RADIONUCLIDES IN THE ATMOSPHERE

4.46. For nuclear installations with low potential hazard, a graded approach may be used. In a graded approach, the following conservative assumptions could be made:

(a)   Conservative source terms are used.

(b)   The representative person is directly downwind of the release at a specified distance close to the source (e.g. at the site boundary) or at the point of peak ground-level concentration for elevated releases.

(c)   Either typical or conservative combinations of wind speed and atmospheric stability (and possibly precipitation) are used. Conservative assumptions might be stable atmospheric conditions with low wind speed for a ground-level release, and unstable conditions that might lead to higher ground-level concentrations closer to the release point for an elevated release.

4.47. If using such conservative assumptions leads to unacceptable results, some of the assumptions might need to be refined. This is elaborated further in Section 10.

5. ANALYSIS OF THE DISPERSION OF RADIONUCLIDES IN SURFACE WATER

5.1. Radionuclides entering surface water are dispersed by general water movements and sedimentation processes. Liquid radioactive releases can be discharged to freshwater, marine or estuarine environments directly. Radionuclides might also be transferred to surface water bodies through atmospheric release followed by deposition on water, or from the ground surface by surface runoff. The potential exposure scenarios and source terms for each accident scenario should be examined based on the safety assessment, including the quantities and relevant physical and chemical characteristics of the releases to the surface water.

5.2. The hydrological dispersion and transfer of radionuclides should be estimated using relevant models, considering the defined hydrological conditions. The output of atmospheric dispersion models may also be used as input for dispersion in surface water if considered significant. As described in Section 7, all exposure pathways should be listed, and the relevant exposure pathways and the representative person should then be identified. Finally, the estimated dose (and, in some cases, a measure of the risk of health effects based on the estimated doses) should be derived and compared with the applicable established criteria. Possible exposure pathways to the representative person through surface water include those associated with drinking water, fisheries, aquatic food, irrigation and recreational activities.

SELECTION OF RELEASE SCENARIOS FOR DISCHARGES AND ACCIDENTAL RELEASES TO SURFACE WATER

5.3. In discharges, radionuclides are directly released into the water body in liquid form or through atmospheric deposition (mainly aerosol) on the surface water. The composition and quantity of the radionuclides should be determined and the chemical and physical forms (e.g. gas, aerosol, liquid) should also be examined to help assess the environmental dispersion of radionuclides.

5.4. In an accidental release, radionuclides might be transferred to surface water bodies either directly or indirectly by deposition. In addition, some of the radionuclides on the ground surface, either due to deposition from atmospheric releases or direct release to the ground, might enter surface water through surface runoff due to precipitation. Such surface runoff should be considered after an accidental release to the ground surface.

SELECTION OF SOURCE PARAMETERS FOR ANALYSIS OF THE DISPERSION OF RADIONUCLIDES IN SURFACE WATER

5.5. The basic source parameters are the same as those for the dispersion of radionuclides in the atmosphere (see paras 4.15–4.22). Additional parameters should be considered in relation to atmospheric deposition, which may be represented by the scenario based models described in paras 4.43–4.45.

5.6. In respect of the source term and receiving water, there should be representative values for all the parameters that affect the dispersion of radionuclides in surface water, including the following:

(a) The radionuclides and the amounts that could be released (e.g. corrosion products, fission products, activation products).
(b) Chemical properties that control the behaviour of radionuclides in surface water, such as adsorption affinity and biological uptake, and chemical form of radionuclides, whether in dissolved or particulate form. Examples of the chemical properties include:
  (i) Major anion and cation concentrations, which control adsorption of radionuclides.
  (ii) Organic content, which is important for the biological uptake of radionuclides by aquaculture.
  (iii) pH, which controls the behaviour of radionuclides in surface water (dissolution affinity of radionuclides).

(iv) Concentration of dissolved oxygen, conductivity and suspended substances.

(v) Salinity, which is important for the marine environment and estuarine areas where fresh water and sea water mix. The water mass characteristics that control the distribution patterns of radionuclides are determined mainly by salinity and temperature.

(c) Physical properties that determine the distribution, dispersion pattern and concentration of radionuclides in the surface water. Examples of physical properties include:

(i) Temperature at multiple depths, which could define the thermocline and vertical distribution pattern of radionuclides in the water and seasonal temperature profile.

(ii) Density, determined by temperature and salinity and varying with water depth. Radionuclides in water parcels with different densities do not exchange; the distribution of radionuclides in surface water elongates within the zone of equal density (isopycnal water parcel).

(iii) Water flow characteristics, which control the dispersion pattern of radionuclides in the surface water.

(iv) Ice cover, especially for lakes.

(v) Sediment load parameters, which control the removal process of radionuclides from surface water to the bottom sediment.

(d) Sedimentation properties:

(i) Distribution coefficient (K_d), which determines the removal of radionuclides from surface water to the bottom sediment;

(ii) Particle size distribution of sediment or surface area of sediment, as indices for adsorption of radionuclides.

Most of these parameters are based on the data collected as background hydrological data listed in Section 3. Additional parameters can be obtained from literature or from more specialized studies in the laboratory or in the field.

MODELS FOR RADIONUCLIDE DISPERSION IN SURFACE WATER

5.7. There are three basic types of models to estimate radionuclide dispersion in surface water:

(a) Numerical models that usually transform the basic equations describing radionuclide dispersion into finite difference or finite element forms.

(b)   Box type models that treat the entire water body or sections of a water body as interconnected homogeneous compartments. These models often include some sediment–radionuclide interactions.

(c)   Analytical models that solve the basic radionuclide transfer equations. Simplifying assumptions can be made regarding water body geometry, flow conditions and dispersion processes to obtain analytical solutions to the governing equations.

5.8.   Other types of models can be used for assessing radionuclide dispersion in surface water systems (e.g. rivers, human-made impoundments, lakes, estuaries, open shores, oceans). Their selection should be based on the quality of results needed for risk assessment.

Radionuclide dispersion in rivers

5.9.   The modelling approach and the level of accuracy of the dispersion modelling depend on the purpose of the model and the accuracy needed for the river under analysis. Simplified models may be developed to represent steady or unsteady flow in one or two dimensions. Detailed modelling typically needs more specific data and more detailed knowledge of the river system. One and/or two dimensional models can be developed in steady or unsteady flow mode using site specific data. For more detailed studies, one or two dimensional models should be used to obtain a preliminary understanding of the behaviour of the hydraulic system and to support a more refined analysis based on three dimensional models.

5.10.   The size and length of the river to be modelled dictates the level of modelling. For example, when the length of the river section is much larger than the width and depth, a one dimensional model should be developed. If the flow path of the water is unknown for some of the events, or if it changes significantly during the event, then a one dimensional model is not appropriate.

5.11.   The analysis objective (i.e. expected type and accuracy of results) should dictate the selection of the appropriate model. Consequently, the source and level of accuracy of the data should be compatible with the selected model. The availability, source and level of accuracy of data should not be used as a basis for developing a model; on the contrary, an appropriate model should be selected to achieve the expected results and the data should be acquired accordingly.

5.12.   Mathematical models, either analytical or numerical, need representative values of the relevant parameters and the boundary conditions, which should be collected with appropriate accuracy either from literature or from site specific studies.

Radionuclide dispersion in impoundments and lakes

5.13. Appropriate models should be selected for radionuclide dispersion in impoundments and lakes. The typical models for dispersion in lakes, along with their advantages and disadvantages for different situations, are as follows:

(a) Box model: The advantages of this model are that the calculation time is short and that long term prediction is possible. Its disadvantages are that the model is not suitable for stratified lakes, cannot represent the heterogeneity within a box and cannot represent the effects of flow changes.
(b) Vertical one dimensional model: The advantages of this model are that the calculation time is short and that long term prediction is possible. Its disadvantages are that the model cannot represent the distribution within a box and that it is difficult to take into account the effects of flow changes (horizontal variation).
(c) Horizontal two dimensional model: The advantages of this model are that the calculation time is shorter than for three dimensional models and that medium term (1–10 year) prediction is possible. The disadvantage of this model is that it is not suitable for stratified lakes.
(d) Vertical two dimensional model: The advantages of this model are that the calculation time is shorter than for three dimensional models, that medium term (1–10 year) prediction is possible and that stratification is represented. A disadvantage is that transverse variation such as horizontal flow is not represented in this model.
(e) Three dimensional model: The advantages of this model are that it can describe local hydrology and water quality characteristics, that it can take into account flow density and drift current and that it can reproduce complex phenomena in the lake. A disadvantage of the model is that a long calculation time is needed.

Radionuclide dispersion in estuaries

5.14. Estuarine regions are connected at one end to a river and at the other end to the sea. Estuary velocity reverses with the tide, and an estuary can contain fresh or saline water, although estuary water is generally less saline than that of the sea. Any radioactive discharge is assumed to occur from one of the estuary banks. The radionuclide concentration at the banks may be assessed using a methodology that is very similar to that used for rivers, but adjustments should be considered to take into account tidal effects, salt wedge and estuarine circulation (i.e. freshwater can flow outwards over saline water flowing inwards, impacting circulation and deposition of radionuclides discharged from banks).

5.15. The hydrodynamics of estuaries are complicated since they are affected by many factors. A number of models are available that can be used for simulating one, two or three dimensional flow, transport and biogeochemical processes in surface water systems. These models can be used to simulate the processes of hydrodynamics, sediment transport, water quality behaviour and eutrophication in one, two and three dimensions. However, such models are usually quite complex and difficult to apply if there are limited observational data for calibration and validation.

Radionuclide dispersion in the open shores of seas and oceans

5.16. Appropriate models for radionuclide dispersion in the open shores of seas and oceans should be selected based on careful consideration of the purpose of the assessment and the level of detail needed for the results. If it is determined that an ocean general circulation model will be deployed, three main types of models could be used to model dispersion of radionuclides in the sea, depending on the vertical coordinate system. These models, along with their advantages and disadvantages for different situations, are as follows:

(a) The z coordinate model, in which the vertical coordinates are perpendicular to gravity. This model is suitable for long term calculations. The z coordinate model utilizes the characteristics of the ocean so that local pressure is expressed as a function of depth by zero-order approximation, which makes implementing the equation of state straightforward. The implementation of bottom topography and drawing of results are also straightforward. This is the most widely used ocean general circulation model because of its versatility. The main disadvantages of this model, however, are that the vertical resolution in shallow seas and near the sea floor tends to be low and that the processes that arise near the coast and the sea floor tend to be poorly reproduced.

(b) The sigma coordinate model, in which the vertical coordinates are the planes along the sea floor. The number of vertical layers to be calculated in shallow water is the same as for deep water. Since the number of vertical grid points is invariable throughout the model domain, sigma models are widely used for coastal ocean simulations. The main disadvantages of this model are that an accurate representation of the horizontal pressure gradient is difficult near steeply sloping bottom topography and that the lateral mixing along the same vertical layer near the continental slope region might lead to the mixing of the shoreward light water and the seaward dense water.

(c)   The isopycnal[7] coordinate model, in which the vertical coordinates of the surfaces are along the isopycnal plane. The development of this class of model is based on the fact that seawater moves along isopycnal surfaces in the interior. Thus, the characteristics of a water mass are well maintained in the ocean interior. Since many theoretical studies of physical oceanography use an isopycnal coordinate framework, isopycnal models have the great advantage of providing good correspondence between theory and numerical models. The main disadvantage of this model is that a surface mixed layer model cannot be incorporated into an isopycnal model.

IDENTIFICATION OF EXPOSURE PATHWAYS FROM SURFACE WATER

5.17. Paragraph 5.27 of GSG-10 [13] lists possible exposure pathways for releases of radionuclides to the atmosphere and surface water in normal operation (typically for nuclear installations such as nuclear power plants). Of these possible exposure pathways, the following are relevant for release to surface water:

— Ingestion of drinking water;
— Ingestion of aquatic food (e.g. freshwater or seawater fish, crustaceans, molluscs);
— External exposure from radionuclides in water and sediments (i.e. from activities on shores, swimming and fishing).

In addition, another pathway to consider is ingestion of food produced using contaminated irrigation water or on land flooded by contaminated rivers, including food produced from animals watered with contaminated water.

5.18. Most accident conditions involve releases to the atmosphere with only indirect releases to surface water. In these situations, assessing only the radiological consequences of the atmospheric release is usually sufficient as these consequences are dominant and any additional impact from indirect releases to surface water is trivial in comparison. Given that the computational effort needed to assess the impact of indirect inputs to surface water is likely to be large (many hundreds of meteorological sequences may need to be considered, taking account of wet and dry deposition in different locations) and that uncertainties in

---

[7] Isopycnals are layers within the ocean that are stratified based on their densities.

the atmospheric source term might be far more significant, consideration should be given to whether such calculations are worthwhile in terms of the end points being determined. Situations where assessment might be worthwhile include atmospheric deposition on reservoirs used for drinking water.

5.19. Accident conditions involving a direct release to surface water should be considered if their likelihood or consequences are such that they could make a significant contribution to the overall risk.

DEFINITION AND COLLECTION OF DATA FOR MODELLING RADIONUCLIDE DISPERSION IN SURFACE WATER

5.20. The data necessary for hydrological analysis come from different sources. The existing national hydrometeorological network usually provides sufficient data. These data, however, should be verified before being used, since their reliability varies depending on the location within the network from where they were collected.

5.21. The data listed in paras 5.22–5.27 are needed for standard calculational models. For advanced models, site specific data may be needed. Typical water bodies in the vicinity of a nuclear installation range from rivers, estuaries and open shores of large lakes, seas and oceans to human-made impoundments.

Modelling in rivers

5.22. The following common characteristics are needed to calculate radionuclide concentrations in a river:

(a)   Cross-section of the river water body;
(b)   Width at the water surface;
(c)   Elevation of the deepest point of the riverbed at a representative cross-section;
(d)   Elevation of the water surface at a representative cross-section;
(e)   Maximum depth;
(f)   Mean depth;
(g)   Local slope of the riverbed;
(h)   Wetted perimeter (wet area per unit length);
(i)   Hydraulic radius;
(j)   Discharge of water;
(k)   Mean flow velocity;
(l)   Friction factor;

(m)   Turbulent diffusion coefficients;
(n)   Coefficient of longitudinal dispersion.

5.23. In addition to these common characteristics, specific characteristics are needed:

(a)   For sediment–water interactions:
      (i)   Suspended solids;
      (ii)   Relative organic carbon content of suspended particles;
      (iii) Sedimentation rate;
      (iv)   Porosity of sediments;
      (v)   Density of sediment particles;
      (vi)   Hydraulic conductivity;
      (vii) Wavelength (horizontal distance) and amplitude of ripples.
(b)   For air–water exchange:
      (i)   Concentration in the atmosphere;
      (ii)   Air–water exchange rate.

Modelling in lakes and impoundments

5.24. The following common characteristics are needed to model the dispersion of radionuclides in a lake or human-made impoundment:

(a)   Geology of the lake or impoundment (e.g. volume–area–elevation curve);
(b)   Seasonal variation of hydrological parameters;
(c)   Longitudinal distance from the release point to a potential receptor location;
(d)   Radionuclide decay constant and decay products;
(e)   Volume of the water body;
(f)   Surface area of the water body;
(g)   Mean depth;
(h)   Discharge rate;
(i)   Total mean concentration;
(j)   Total external input of the radionuclide per unit volume and time;
(k)   Specific river input and output rates;
(l)   Specific rate of constant of air–water exchange;
(m)   Specific removal rate by particle settling;
(n)   Specific exchange rate between water and the sediment column;
(o)   Specific removal rate by water exchange with adjacent box or boxes;
(p)   Specific chemical transformation rate;
(q)   Specific photochemical transformation rate;
(r)   Specific biological transformation rate;
(s)   Mean total concentration in inflow;

(t)   Atmospheric concentration;
(u)   Apparent sediment concentration for input;
(v)   Total concentration in adjacent box;
(w)   Dissolved fraction of the chemical species;
(x)   Solid–water distribution coefficient;
(y)   Solid–water phase ratio.

Modelling in estuaries

5.25. The following characteristics, in addition to those listed for lakes and impoundments in para. 5.24, are needed to model the dispersion of radionuclides in an estuary:

(a)   Estuary width;
(b)   Estuary flow depth;
(c)   River width under a mean annual river flow rate upstream of the tidal flow area;
(d)   Tidal period.

Modelling in open shores of seas and oceans

5.26. All oceanic phenomena affecting dispersion should be considered. The representative physical factors for developing the oceanic models in terms of their time and spatial scales are given in Table 1. However, using a general circulation model that includes marine areas in the site evaluation does not necessarily provide the information needed for the evaluation.

5.27. The ocean general circulation model considers wind-driven circulation and thermohaline circulation to represent the global scale. Global models are typically used as a boundary condition for the regional model that represents the target ocean. The regional model should represent the relevant physical oceanographic phenomena, such as mesoscale eddies, swells and wind waves, in order to represent the topography and ocean currents specific to the target area. A high resolution model with a grid size of a few kilometres is often used near the coast, and a low resolution model with a grid size of 10–100 km is used in the open ocean.

TABLE 1. REPRESENTATIVE PHYSICAL FACTORS FOR DEVELOPING OCEANIC MODELS

Representative physical factor	Timescale	Spatial scale
Wind waves	1–10 s	1–10 m
Microstructure turbulence	1 s to 1 min	1 cm to 1 m
Boundary layer turbulence	1 min to 1 day	10 cm to 100 m
Swell	1 s to 1 min	100 m
Internal gravity waves	1 hour to 1 day	100 m to 10 km
Submesoscale currents	1 hour to 1 month	100 m to 10 km
Mesoscale eddies	1 day to 1 year	1–100 km
Tides	1 hour to 1 day	1 000–10 000 km
Wind-driven circulation	1 month to 100 years	100–1 000 km
Thermohaline circulation	100–1 000 years	1 000–10 000 km

CALIBRATION OF MODELS AND SENSITIVITY STUDY OF THE ANALYSIS OF RADIONUCLIDE DISPERSION IN SURFACE WATER

5.28. The results from a calculational model should be compared with laboratory data or field data for a specific site. Such validation usually has a limited range of applicability, which should be determined with a full understanding of the model. The model should be calibrated by comparing it with the actual environmental monitoring data set. Calibration of the model can be achieved by selecting some representative (or tuning) parameters measured at the site. Tracer tests can also be conducted in rivers, impoundments, lakes and estuaries, and the dispersion of the tracers can be observed. Tracer tests can also provide determination of some transport parameters such as diffusion and dispersion. It should be verified, for example, that the errors and uncertainties in the model output values are within the error range of the actual observed values.

5.29. As with the atmospheric dispersion model, the assessment involves many assumptions and uncertainties, so a sensitivity study should be performed to assess the sensitivity of the overall results to the assumptions and parameter values. A sensitivity study should include the following:

(a) Hydrological data to be used;
(b) Source term assumptions including radionuclide activity released, potential water depth of release and surface deposition associated with atmospheric fallout;

(c)   Representative assumptions (e.g. age group, residence, food consumption, water consumption);

(d)   Assumptions about parameter values such as deposition rates;

(e)   Assumptions about the measures to be applied.

SCENARIO BASED SIMULATION OF RADIONUCLIDE DISPERSION IN SURFACE WATER

5.30. The scenario for discharge should assume that radionuclides are released to surface water at a constant release rate and that the release continues for the lifetime of the installation. The calibrated model can be run for various scenarios of accidental releases. The scenarios may include different locations of input at various modes of injections (with various amounts of releases, concentrations of radionuclides and time and/or duration of the release). Scenarios should be developed based on the design of the installation as well as the hydrographical setting of the site. Surface water deposition associated with short term planned releases to the atmosphere, such as those that occur during maintenance, can also be simulated.

5.31. In addition, a series of radiological effects can be simulated by reproducing the diffusion of radionuclides in surface water through calculations based on the release of various types and amounts of radionuclides in multiple accidental releases and the corresponding sample times from hydrological data sets.

GRADED APPROACH TO ASSESSING THE DISPERSION OF RADIONUCLIDES IN SURFACE WATER

5.32. The level of complexity of a dispersion model for radionuclides in surface water should be chosen primarily according to the magnitude of the installation's hazard category (see para. 10.5) and the complexity of the hydrological environment. In particular, before developing a detailed model, it is useful to simplify the site characteristics and consider conservative dispersion mechanisms.

5.33. When assessing a river, its size, length and profile, as well as local flow conditions at the discharge point, should be taken into account when determining the level of model complexity. For example, if the river's reach length is much greater than its width or depth, a one dimensional model may be used since complete mixing can be assumed. If the width of a river is so great that discharged water flows over long stretches of the watercourse along one bank and only mixes slowly over the total width, a one dimensional model is not appropriate and a more sophisticated model should be used. The same is true if discharged water flows via a tributary into another river. When assessing the lateral mixing in the river, different flow conditions, especially low water, should be taken into account. If the water flow path is unknown for certain events or if it changes significantly during an event, a one dimensional model is not appropriate and a more sophisticated model should be used.

5.34. The basic flow phenomena in lakes and human-made impoundments are the flow due to the inflow and outflow of rivers, and the wind-driven flow. The presence or absence of vertical stratification associated with seasonal changes in air and water temperatures is also a criterion for determining whether the model can be simplified. With regard to the spatial scale, a low-dimensional model may be selected when a rough scale such as the average water quality in the lake is sufficient. With regard to the timescale, if the long term variation over a year or more needs to be determined, a low-dimensional model should be considered because a high-dimensional model might not be practical. If a short term phenomenon such as runoff or storm surge needs to be determined, a high-dimensional model would be more appropriate to achieve sufficient accuracy. Section 10 provides further recommendations on the application of a graded approach.

5.35. In the flow field in the ocean, various processes should be considered, such as three dimensional modelling of water mixing associated with temperature, salinity, density, tidal fluctuations, freshwater supply from rivers, the influence of strong currents due to thermohaline circulation in the open ocean, and the presence or absence of eddies. In coastal areas, various processes can be applied to simplify the model depending on the features of the region, such as the presence or absence of large rivers, the seasonal development of vertical stratification and the influence of tidal currents.

6. ANALYSIS OF THE DISPERSION OF RADIONUCLIDES IN GROUNDWATER

6.1. The objectives of conducting a hydrogeological study at a nuclear installation site and in the vicinity of the site are to determine the following:

(a) The estimated concentration of radioactive material in groundwater at the nearest point in the region where groundwater is drawn for human consumption;
(b) The transport paths and travel times for radioactive material to reach the source of consumption from the point of release;
(c) The transport capacity of the surface flow, interflow and groundwater recharge;
(d) The susceptibility to contamination of an aquifer or of aquifers at different levels;
(e) The temporal and spatial distributions of the concentrations of radioactive material in the groundwater resulting from discharges and/or accidental releases from the nuclear installation.

6.2. A graded approach can be applied to the scope of the hydrogeological study in accordance with the hazard category of the installation.

6.3. The hydrosphere is a major medium by which radioactive material that is released from a nuclear installation via discharges or accidental releases could be transferred into the environment and to locations where water is used by or for the population. The dispersion of radionuclides in groundwater is very slow compared with their dispersion in surface water (except in karst topography).

6.4. A detailed investigation of the hydrogeology in the region should be performed. Calculations of the transport and concentrations of radionuclides should be made to show whether the radiological consequences of potential releases of radioactive material to the groundwater (see para. 6.12) are acceptable.

6.5. The results of the hydrogeological investigation should be used for the following purposes:

(a) To confirm the suitability of the site;
(b) To select and calibrate an appropriate flow and transport model for the site;
(c) To establish limits for radioactive discharges to pathways that ultimately reach the groundwater;

(d)   To assess the radiological consequences of releases;

(e)   To assist in evaluating the feasibility of planning effective emergency response actions;

(f)   To develop a monitoring programme and a sampling strategy for use in normal operating conditions and also in the event of an accidental radioactive release.

6.6.   The information necessary to perform dose assessment relating to exposure pathways in the hydrogeological system includes the following:

(a)   The source term from radioactive material released to the groundwater system;

(b)   Hydrological, physical, physicochemical and biological characteristics governing the dispersion, diffusion and retention of radioactive material;

(c)   Relevant food chains leading to humans.

6.7.   The direction of groundwater movement and of radionuclide transport in isotropic media is orthogonal to the contours at the hydraulic head. In this case, the standard calculational models may be applied. If the aquifers are strongly anisotropic, however, and the water and transported effluents can move over a limited domain through fractures and/or karstic conduits, most calculational models are not valid. In this case, field studies including tracer tests may be necessary and should be considered. The level of complexity of the model should primarily be selected on the basis of the level of risk of the installation and complexity of the hydrogeological configuration.

6.8.   The objectives outlined in para. 6.1 may be achieved primarily by mathematical models that produce groundwater flow velocity vectors in the flow domain. These models should then be coupled with transport models to assess the spatial and temporal variations in the concentrations of radionuclides. Computer codes that are capable of evaluation of groundwater by combining flow and transport can be used.

6.9.   The calculational model should also be selected on the basis of the objective of the study. Considering the objectives listed in para. 6.1, preference should be given to process based deterministic models. The models selected should be suitable for simulating the transport, dispersion, dilution and accumulation of radionuclides, and their decay or other removal mechanisms, as necessary. The mode of the releases expected during normal operation of the installation as well as potential exposures should be taken into account (see also GSG-9 [20]).

6.10. When it is appropriate to use analytical models, a detailed analysis of appropriateness in terms of the boundary conditions and assumptions that satisfy the physical conditions at the study site should be conducted. Consequently, the analytical model used should be validated for each specific application.

6.11. Considering their characteristics, analytical models for groundwater flow and radionuclide transport should be applied as an initial prediction because, in some cases, they may involve a high level of simplification of the real system. Additionally, the assumptions in these models limit their application to relatively simple systems. Therefore, careful consideration should be given in selecting analytical models for groundwater flow and radionuclide transport.

SELECTION OF RELEASE SCENARIOS FROM DISCHARGES AND POTENTIAL EXPOSURE RELEASES TO GROUNDWATER

6.12. A release of radioactive substances from a nuclear installation might contaminate the groundwater system in the region either directly or indirectly, via soil, atmospheric fallout or surface water, in the following ways:

(a) Indirect discharge to groundwater through seepage and infiltration of surface water that has been contaminated by radioactive substances discharged from the nuclear installation;
(b) Infiltration into groundwater of radioactive liquids from a storage tank or reservoir;
(c) Infiltration into groundwater of any airborne radioactive material deposited on the ground surface or on surface water;
(d) Direct release from a nuclear installation as a result of an accident.

6.13. The potential for indirect contamination of surface water and possible contamination of groundwater from the surface should be assessed.

6.14. The protection of aquifers from accidents should be considered in the safety analysis for postulated accident conditions, taking into consideration geological barriers. As a result of safety analysis, the building of a protective barrier might be necessary.

SELECTION OF SOURCE PARAMETERS FOR ANALYSIS OF THE DISPERSION OF RADIONUCLIDES IN GROUNDWATER

6.15. The following properties and parameters should be estimated for radioactive releases:

(a) Radioactivity:
  (i)   Rate of release of each important radionuclide;
  (ii) Location, duration and concentration of the release.
(b) Chemical properties, including the following:
  (i)   Concentrations of important anions and cations, and their oxidation states and complexing states (e.g. Ca^{2+}, K^+, Mg^{2+}, Na^+, NH_4^+, HCO_3^-, Cl^-, SO_4^{2-}, NO_2^-, NO_3^-, PO_4^-);
  (ii) Organic content;
  (iii) pH and Eh value (redox potential);
  (iv) CO_2 partial pressures;
  (v)   Chelating agents;
  (vi) Concentration of dissolved oxygen, and electrical conductivity and concentrations of associated pollutants.
(c) Physical properties of the liquid effluents discharged, including:
  (i)   Temperature;
  (ii) Density;
  (iii) Loads and granulometry of suspended solids.
(d) Flow rates for continuous discharges, or volume and frequency for batch discharges.
(e) Variation of the source term over the duration of the discharge.
(f) Geometry and mechanics of discharges.
(g) Sorption/desorption characteristics of the specific radionuclide onto sediments and/or rock matrix.
(h) Distribution coefficient(s) (K_d) between the liquid phase and the solid phase.
(i) Physical, chemical and geological factors that can influence K_d.
(j) Diffusion parameters.

CONCEPTUAL MODEL DEVELOPMENT

6.16. A variety of models and data are necessary to predict the dispersion of radionuclides through environmental media and transfer to the representative person. The processes that are more relevant to dose estimation should be identified and a conceptual model should be elaborated in the form of a representation that captures the key elements or components of a complex system, such as the relationship between the released radionuclides and the environment. The conceptual model should represent the identified relevant exposure pathways (see para. 2.9).

6.17. A conceptual model that provides a working description of the characteristics and dynamics of the hydrogeological system is essential for any analysis of the water flow and dispersion of radionuclides. It should be regarded as the fundamental step of hydrogeological assessment for a nuclear installation site.

6.18. Conceptualization and characterization of the hydrogeological system are essential and form the most important part of the predictive flow and dispersion modelling. By nature, conceptual models are a simplification of a real system. However, the degree of simplification should be decided in accordance with the type of nuclear installation and the stage of reporting. (See Section 10 for more details on a graded approach.)

6.19. Inadequate conceptualization is one of the main sources of uncertainty and might result in unreliable models for the dispersion of radionuclides. Inadequate consideration of spatial variations of hydrogeological parameters might also adversely affect the results. Simple hydrogeological models that might not produce a conservative assessment of the system behaviour should be used with caution.

6.20. It is possible to construct a preliminary hydrogeological conceptual model for a nuclear installation site on the basis of geological and hydrological information available for the site itself and/or its near vicinity. Properties of similar geological materials elsewhere and generic data from similar geographical and geological regions can also be used in preliminary conceptualization. However, each site is unique in its hydrogeological features, properties and behaviour, so it is not possible to fully represent the hydrogeological setting using a generic conceptual model. Therefore, site specific conceptualization and characterization should be the ultimate objective of the hydrogeological data acquisition.

6.21. To decide the extent of the study area, first the hydrogeological domain to which the nuclear installation site belongs should be defined. A model area should then be determined for hydrogeological conceptualization and characterization. The conceptual model should extend to natural boundaries (e.g. topography such as topographical divide, geological structure or lithological contact, or surface water features such as streams, rivers or lakes). The model should also consider the extent of the potential impact of stress generated at the site. To reduce the impact of boundary conditions on the model, the extent of the hydrogeological domain to be studied should be larger than the model domain.

6.22. Alternative conceptual models can be constructed based on the available data, and any reasonable alternative conceptualizations should be evaluated. Further studies should be performed in order to reduce model bias and uncertainty and thus ensure the most appropriate and/or representative conceptual model.

6.23. An iterative approach should be used in the process of construction of a hydrogeological conceptual model. The preliminary conceptual model should be tested by an appropriate mathematical model (see paras 6.24–6.33) using the monitored data and refined until improvements in the predictive capability of the model are, practically, not necessary. Prediction refers to mainly the flow (head distribution). Monitoring the temporal variation of groundwater levels provides the time series needed to check the capability of the model by comparing the predicted head with the monitored data.

DEFINITION AND COLLECTION OF DATA FOR MODELLING RADIONUCLIDE DISPERSION IN GROUNDWATER

6.24. Hydrogeological investigation in the framework of site evaluation for a nuclear installation involves regional and local investigations using comparatively standard hydrogeological mapping, surface geophysical surveys and borehole drilling programmes for hydrogeological characterization studies such as packer tests, single well tests, pumping tests, geophysical studies and tracer tests.

6.25. Both regional and local information should be collected to identify the hydrogeological system and determine whether preferential flow paths are present. The information to be collected should include the following:

(a) Meteorological data: In regions where precipitation (e.g. rain, snow) makes a substantial contribution to groundwater, long term meteorological data on annual, monthly and, if available, daily precipitation and on corresponding air temperature (to calculate potential evapotranspiration) should be analysed for as long a period as there are data available. The average precipitation should be calculated using appropriate interpolation techniques from precipitation data recorded at meteorological stations in and around the watershed where the installation is situated. The effect of topography may need to be considered where there is a large difference in elevation. Meteorological data analyses should also be performed for the groundwater recharge at an acceptable level of certainty. Alternatively, tracers (chemical or isotopic) of the water cycle as well as satellite technologies could be introduced to calculate groundwater recharge.

(b) Surface runoff: Another component of the water balance should be either estimated or measured at the outlet of the basin. Long term records of flow of the stream draining the basin may be used to assess the surface runoff. If there is no flow record, empirical relationships or satellite technologies may be used to estimate the surface runoff. The construction of weirs or flumes to measure the flow rate for at least one water year[8] should be considered. This would help with making adjustments to the empirical assessment of the runoff coefficient and the surface runoff. The baseflow component of the measured streamflow may need to be considered and calculated.

(c) Discharge data of significant springs: Springs with significant discharge should be identified, including their type, and their discharge should be measured on at least a monthly basis for a minimum of one water year.

(d) Surface drainage system or standing water body: Any surface drainage system or standing water body accessible from a potential release point in an accident should be identified. Areas from which contaminated surface water can directly enter an aquifer (e.g. sinkholes) should be determined. In karstic areas, the recharge to the aquifer may be either concentrated or diffused. Concentrated recharge implies that the source of the recharge is surface water directly above the aquifer. Diffused aquifer recharge implies that the recharge source is not directly above, but rather that the aquifer recharges through hydraulic communication at depth. The karstic features (e.g. ponors, swallow holes, sinkholes) that allow direct input to the aquifer should be mapped and considered in characterization of the groundwater system. The relevant hydrogeological information for surface or near surface discharges includes information on soil moisture properties, infiltration rates, configuration of unsaturated zones and chemical retention properties under unsaturated conditions. In addition, records of level fluctuation of standing water bodies such as lakes and wetlands are needed for a complete water balance calculation. Bathymetry should be mapped for establishing the elevation–area–volume relationship.

(e) Description and mapping of major hydrogeological units: Data should be obtained on the various types of geological formations in the region and their stratigraphical distribution in order to characterize the regional system. The hydrostratigraphical units should be described on the basis of hydrogeological properties of the lithological units in the region. For consideration of the transport potential of seepage and groundwater in the region of the site, data on types of aquifers, aquitards and aquicludes, their

---

[8] A water year is a 12-month period over which precipitation totals are measured, not normally corresponding to a calendar year. Normally the year begins on the first day of the month in which a dry period ends and precipitation increases.

interconnections and the flow velocities and mean transit times should be investigated. The extent and thickness of major hydrostratigraphical units, in particular of the aquifer units, should be mapped and depicted on cross-sections. Three dimensional visualization should be provided. Karstic features such as sinkholes, dolines, poljes and similar closed depressions, caves and underground rivers should be mapped.

(f) Hydraulic head distribution: Potentiometric maps should be prepared for each aquifer (if the flow domain is a multiaquifer system), for at least one dry and one wet period. The potentiometric map should be produced from the groundwater levels measured in a sufficient number of piezometers. Heterogeneity should be considered in deciding on the number and locations of the piezometers. Such data permit the regional flow pattern and its relation to the local flow pattern of seepage and groundwater to be characterized. Dye tracing tests should be designed and conducted in karstic aquifers to delineate the groundwater catchment area and assess the direction and velocity of groundwater flow.

(g) Description of natural recharge and discharge areas: Potentiometric maps can also be used to delineate recharge and discharge areas, and to define hydraulic boundaries and boundary conditions of the flow domain. Environmental isotopes (stable and radioactive) should be considered a useful tool in assessment of recharge–discharge relationships. Stable isotope characteristics of local and regional precipitation should be obtained to establish the relationship between elevation and ^{18}O. This relationship can be obtained by analysing seasonal springs issuing at different altitudes.

(h) Ages, transit time or mean turnover times of groundwater: Artificial or environmental tracers such as tritium, the ratio of helium to tritium (where tritium is close to the natural background), or other appropriate tracers should be used to obtain the average apparent age, transit time and turnover time of groundwater. In complex systems, a vertical profile of groundwater age should be determined. Environmental isotopes and hydraulic heads should be used to investigate interconnections between aquifers, and interactions between groundwater and surface water.

(i) Hydrochemical data: Water samples from groundwater (e.g. springs, wells) and surface water bodies should be collected properly and analysed for major ion content at least on a seasonal basis. In situ measurements of temperature, electrical conductivity, pH, redox potential and dissolved oxygen should accompany the sampling.

(j) Hydraulic characteristics and transport parameters: A sufficient number of laboratory and/or field tests should be performed to obtain representative values of the hydraulic characteristics of each aquifer material, such as porosity, hydraulic conductivity, transmissivity, storativity (i.e. storage coefficient), specific yield, bulk density of aquifer material and dispersivity (i.e. hydrodynamic dispersion coefficient). Batch, column experiments and/or in situ tracer tests should be performed to determine the sorption characteristics (e.g. distribution coefficients) for radionuclides of interest.

MODELS FOR RADIONUCLIDE DISPERSION IN GROUNDWATER

6.26. Interpretive (also known as informative) and predictive models can be used to model radionuclide dispersion in groundwater. Interpretive models are used to obtain a thorough understanding of the hydrogeological system dynamics. They help to construct and to test the hydrogeological conceptual model of the site. Interpretive models do not necessarily need to be calibrated but predictive models do. Calibration mainly refers to flow models, with the assumption that a calibrated flow model will provide a substantial basis for a useful transport model. Calibration of transport models can be achieved by tracer tests, considering advective and dispersive transport only.

6.27. Activity concentrations in the subsurface environment resulting from the postulated discharge of radioactive material should be estimated by means of mathematical models. A number of models have been developed to calculate the dispersion and retention of radionuclides released into groundwater. Standard calculational models are generally satisfactory and should be used in most cases. The complexity of the model chosen should reflect the complexity of the hydrogeological system at a particular site. The objective of modelling should also be taken into consideration during the selection of the model. Further recommendations on the selection of the appropriate level of complexity of the model to be used are provided in Section 10.

6.28. Two possible approaches can be taken to the use of models and data for the assessment of radionuclide dispersion in groundwater. A generic and simple methodology can be followed, which takes account of the dilution, dispersion and transfer of radioactive material in the environment with conservative assumptions. Alternatively, a specific, more detailed methodology can be followed, using site specific data to estimate activity concentrations in different environmental media, with more realistic assumptions. In some situations, a combination of generic models with site specific data could also be suitable for the assessment. In all cases, the models selected should be suitable for estimating the spatial distribution and temporal variation of activity concentrations in the environment. The complexity of the model used should be commensurate with the likely level of environmental impact from the installation.

6.29. The models that can be used for nuclear installations are diverse and can be categorized according to the problem being addressed. Deterministic models and stochastic models are the main two categories commonly used by States. The most appropriate approach should be chosen on the basis of the hydrogeological setting (conceptualization and characterization), and the level of accuracy sought at the reporting stage of interest (e.g. site characterization stage, construction stage, operation stage, closure of the installation).

6.30. Stochastic models may be used to consider strong heterogeneity and occurrence of preferential flow paths. Stochastic models are used when there are significant sources of uncertainty. Occurrence of preferential flow paths may also be considered as one of the sources of heterogeneity. Geostatistical methods are useful in producing the spatial variability of parameters that is needed in a stochastic approach. Monte Carlo simulation is the most commonly applied stochastic approach to predict groundwater flow and radionuclide transport on the basis of geostatistical inputs. This approach can also be used to reduce and quantify the predicted uncertainty.

6.31. Deterministic models can be subcategorized as lumped (also referred to as black box or grey box) models or distributed parameter (also referred to as process based) models. Mathematical (partial differential) equations simulating groundwater flow and solute (radionuclide) transport are the most used distributed parameter models. These equations are solved either analytically (i.e. with exact solutions) or numerically (i.e. with approximate solutions, commonly known as mathematical models).

6.32. Analytical models are solutions that satisfy certain geometry and specific boundary conditions of the flow domain, and are generally limited in their consideration of heterogeneity and anisotropy. Significant uncertainties may be associated with these models when the assumptions are not totally satisfied at the site under study. When the hydrogeological conceptual model partly meets the boundary conditions and the assumptions of the selected analytical model, they can be used as a first approximation and the result should be evaluated with caution.

6.33. Numerical flow and transport models can be applied with different levels of simplification. Flow and solute transport phenomena in the subsurface environment might involve various processes. In particular, the transport models are commonly known by the process or processes involved, such as advective, dispersive, sorptive, reactive or radioactive, or a combination of some or all of these processes.

6.34. The vadose zone in general has a retarding and attenuating role in contaminant fate and transport behaviour. However, characterizing the transport properties of the vadose zone can be difficult. Ignoring the effects of the vadose zone in modelling can lead to higher concentrations and faster movement of the contaminants in saturated zone groundwater. Such a modelling decision allows for making conservative predictions at the first stage of site evaluation. Conservative assumptions and sensitivity tests can help to inform the need to model the vadose zone. If the site is determined to be acceptable without vadose zone modelling, characterization of the vadose zone may be omitted. Experience has shown that the following affect the level of complexity of the hydrogeological model:

(a)   Ignoring the role of the unsaturated zone;
(b)   Considering a conservative contaminant;
(c)   Assuming a homogeneous and isotropic flow domain.

6.35. Modelling should start with the simplest model — advective — which assumes that the dispersion is governed only by the mean velocity of groundwater flow. Therefore, descriptions of dispersion parameters and variables are not needed.

6.36. In order to use more complicated models (e.g. combining all processes), more hydraulic and dispersion parameters need to be determined, such as dispersivity, diffusion coefficients, distribution coefficients, kinetic reaction rates and half-lives. Recommendations on the application of a graded approach for different reporting stages are provided in Section 10 and in the Appendix.

IDENTIFICATION OF EXPOSURE PATHWAYS FROM GROUNDWATER

6.37. The following can serve as the starting point for possible exposure pathways from groundwater during normal operation of nuclear installations:

(a)   Boreholes, wells and galleries used to abstract water for drinking;
(b)   Springs captured for drinking water;
(c)   Groundwater used for agriculture;
(d)   Groundwater base flow to streams, rivers, lakes or wetlands (that can lead to internal exposure through ingestion of drinking water and/or aquatic food such as fish, crustaceans and molluscs);
(e)   Groundwater flow into the sea and estuaries (that can lead to internal exposure through ingestion of aquatic food and external exposure through activities such as swimming and fishing).

CALIBRATION OF MODELS AND SENSITIVITY STUDY OF THE ANALYSIS OF RADIONUCLIDE DISPERSION IN GROUNDWATER

6.38. Models always have limitations because they are simplifications of the complex real world and do not provide a unique representation of reality. However, properly constructed models can provide reliable results within the uncertainty limits. Therefore, the level of uncertainty of a model should be evaluated and reported.

6.39. The calibration of a model provides the means to test and/or compare the selected conceptual models. Calibration involves observation of the actual site conditions and thus the monitoring of data sets. To ensure that the model simulates the real system with an acceptable degree of error and uncertainty, the calibration should be done for both steady state and transient conditions.

6.40. Consideration should be given to uncertainties that might arise from (a) deficiencies in understanding and conceptualization of the hydrogeological system; (b) spatial and temporal variations in variables and parameters; and (c) definition of the boundary conditions of the flow domain. Different types of uncertainty (e.g. conceptual model uncertainty, numerical model uncertainty, parameter uncertainty, exposure scenario uncertainty, exposure parameter uncertainty) should be considered.

6.41. A sensitivity study should be conducted to identify the parameters and locations to which the system behaviour is sensitive. Performing additional site characterization to better estimate parameters at these locations reduces model uncertainty. Further monitoring should be performed where the system is most sensitive to model parameters.

SCENARIO BASED SIMULATION OF RADIONUCLIDE DISPERSION IN GROUNDWATER

6.42. To achieve the objectives described in para. 6.1, a predictive model should be run to simulate different scenarios. A properly constructed and calibrated model provides a tool to forecast the response of the groundwater system to future conditions.

6.43. Two sources of uncertainty should be considered in scenario based simulations[9]. The model itself is one source of uncertainty (see para. 6.38) and the other is associated with the scenario. The accuracy of specification of the future conditions should be considered as a significant source of uncertainty in the forecast.

6.44. Primarily, simulation under normal conditions (continuous discharge) should be run for different scenarios. Scenarios should be based on the expected future changes in natural conditions and on the design of the installation. Changes in the meteorological and hydrological conditions during the lifetime of the installation and the release of radionuclides during normal operation should be simulated for a period covering at least the lifetime of the installation. Changes in meteorological parameters such as precipitation, temperature and land use, which affect surface runoff and evapotranspiration, should be taken into account. The exposure pathways defined in para. 6.37 should also be considered.

6.45. Similarly, different scenarios defining possible types and locations of an accidental release of radionuclides should also be simulated to forecast the pathways, distribution of concentration, activity and velocity of the radionuclides in the groundwater system. Interactions with surface water bodies should be considered, where applicable.

GRADED APPROACH TO ASSESSING THE DISPERSION OF
RADIONUCLIDES IN GROUNDWATER

6.46. Detailed recommendations for determining the most appropriate level of complexity for modelling radionuclide dispersion in groundwater are provided in the Appendix.

---

[9] In a scenario based simulation, the calibrated model is run to simulate different plausible release scenarios for discharges and accidental releases.

7. ASSESSMENT OF THE OVERALL RADIOLOGICAL IMPACT OF A NUCLEAR INSTALLATION

SUMMARY OF THE NUCLEAR INSTALLATION SITE CHARACTERISTICS USED FOR RADIOLOGICAL IMPACT ASSESSMENT

7.1. Recommendations on the characteristics of a nuclear installation site that form the basis for radiological impact assessment are provided in Section 3. These characteristics primarily relate to meteorological and hydrological conditions, topography, population distribution and habits, land and water use, natural background radioactivity and food production and consumption in the vicinity of the site. These site characteristics and where they are used in radiological impact assessments for a nuclear installation are shown in Table 2 and Fig. 1.

7.2. External hazards at the site might affect the installation and this should be considered in the analyses for determining the types of safety features that are incorporated into the design of a nuclear installation and the frequency of potential accident scenarios during the operation of the installation. Therefore, these design characteristics also affect the radiological impact of nuclear installations either by altering the source terms (i.e. quantities, physical and chemical form and timing of radionuclides released to the environment during an accident) or by changing the frequency of potential accident scenarios. Source terms have a strong influence on the potential doses from individual accident scenarios and the frequency of accidents has a direct effect on the total radiological risk from a nuclear installation (see paras 7.27–7.30).

TABLE 2. OVERVIEW OF SITE CHARACTERISTICS AND THEIR USE IN RADIOLOGICAL IMPACT ASSESSMENTS FOR NUCLEAR INSTALLATIONS

Site characteristic	Use in radiological impact assessment				
	Atmospheric dispersion	Surface water dispersion	Groundwater dispersion	Public dose estimation	Biota dose estimation
Meteorology:					
Wind speed and direction, temperature, humidity and atmospheric stability	X		X		
Precipitation	X	X	X		
Surface water hydrology:					
Physical characteristics		X			
Flow and interconnections		X			
Sedimentation		X			
Interaction with groundwater		X	X		
Groundwater hydrology:					
Site hydrogeology (i.e. description and mapping of major hydrogeological units)			X		
Groundwater flow, including recharge and discharge			X		
Hydraulic head distribution and hydraulic characteristics[a] of aquifers		X	X		
Surface features, topography	X	X	X		

TABLE 2. OVERVIEW OF SITE CHARACTERISTICS AND THEIR USE IN RADIOLOGICAL IMPACT ASSESSMENTS FOR NUCLEAR INSTALLATIONS (cont.)

Site characteristic	Use in radiological impact assessment				
	Atmospheric dispersion	Surface water dispersion	Groundwater dispersion	Public dose estimation	Biota dose estimation
Population distribution and habits				X	
Habitats for biota[b]					X
Land use	X	X	X	
Water use	X	X	X	
Food production and consumption				X	

[a] Examples include porosity, conductivity, transmissivity and dispersivity.
[b] See annex 1 to GSG-10 [13] for an example of a generic methodology.

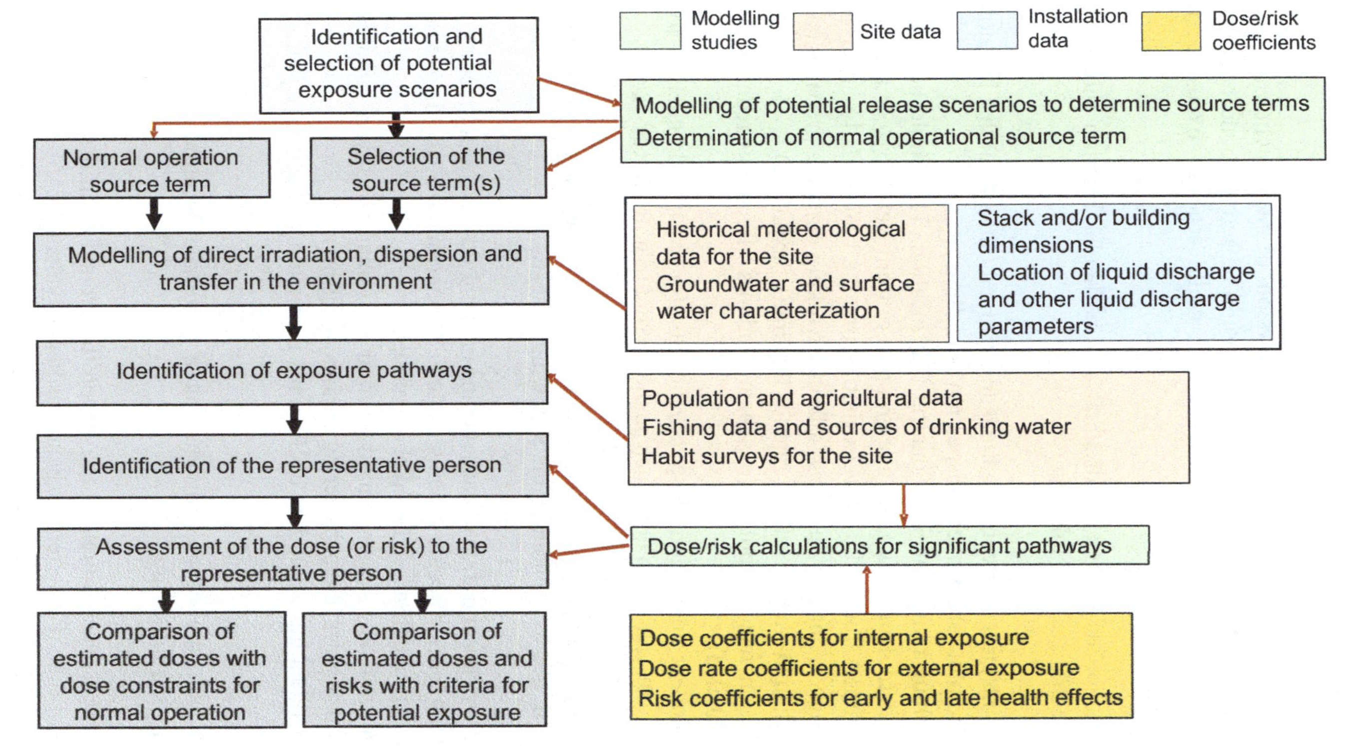

FIG. 1. Data used and modelling performed at the various stages of the radiological environmental impact assessment for operational states and accident conditions.

COMPILATION OF RADIOLOGICAL IMPACT ASSESSMENT
RESULTS

7.3. The end points for analysis of the dispersion of radionuclides in the atmosphere, groundwater and/or surface water should be activity concentrations of radionuclides in environmental media that could lead to exposure of people and biota.

7.4. The inhalation dose for an individual at a particular location should be determined by combining the following:

(a)   The time-integrated air concentrations for each radionuclide at that location derived from the atmospheric dispersion modelling.
(b)   The breathing rate of an individual at that location. This could be taken from standard data for a given age group (e.g. from Ref. [25]), or it could be determined from habit surveys that record the typical daily hours of performing various activities (e.g. a farmer working outdoors probably has a higher than average breathing rate for the time of exposure).
(c)   Any location factors applied (e.g. being indoors effectively reduces the activity inhaled).
(d)   Age-dependent inhalation dose coefficients for each radionuclide for effective dose and organ dose such as the thyroid (e.g. from Ref. [26]).

7.5. The ingestion dose of an individual consumer of a particular food type should be determined by combining the following:

(a)   The activity concentration of each nuclide in that food type (as a function of time if long term exposure is being evaluated) determined by the dispersion and transfer modelling and food chain models;
(b)   The ingestion rate for an individual food consumer, usually determined from habit surveys and considering various age groups;
(c)   Age-dependent ingestion dose coefficients for each radionuclide (e.g. from Ref. [26]).

7.6. External exposure of an individual at a particular location should be calculated using the following:

(a)   The activity concentrations in the air, on the ground or in the water (as a function of time if long term exposure is being evaluated);
(b)   The time the individual is exposed, which is determined from habit data;

(c)   The geometry of the exposure and distance from sources of radioactivity
       (e.g. deposited activity on the ground, a plume in the air);

(d)   Any factors applied for shielding (e.g. being indoors);

(e)   The dose rate coefficients for each radionuclide (e.g. from Ref. [27], which
       provide values for typical environmental geometries and shielding situations
       for exposure from air, soil and water).

7.7.   The total dose for any given individual is the sum of all dose contributions
listed in paras 7.4–7.6 (i.e. inhalation, ingestion and external exposure).

7.8.   To meet any regulatory criteria for individual dose or risk, a representative
person should be identified (see paras 7.14–7.17).

7.9.   For normal operation of a nuclear installation, a criterion for a single
source or site can be a dose constraint that is a fraction of the dose limit for
public exposure, which is an effective dose of 1 mSv in a year (see para. III.3 of
GSR Part 3 [4]).

7.10. The radiological impact assessment should consider the whole lifetime
of the nuclear installation so that the effects of accumulation of longer lived
radionuclides in the environment can be assessed.

7.11. For all accident conditions considered, the individual doses from the
various release routes (i.e. atmosphere, surface water, groundwater) and exposure
pathways should be summed for comparison with either reference levels or dose
constraints and dose limits, as appropriate.

7.12. For public exposure from accidental releases from nuclear power plants
or other types of nuclear installation, it may be sufficient to only consider
atmospheric releases, since the atmosphere is likely to be the release route for
most accident situations unless there are accident scenarios involving a direct
release to surface water or groundwater. Compared with an atmospheric release,
the impact from a release to surface water or groundwater is less immediate:
dilution in large bodies of water can occur and the water provides some shielding
(see para. 2.28), reducing the exposure of any people in the vicinity. Where there
is any uncertainty regarding the proportion of a release to the atmosphere and to
water bodies, the conservative assumption would usually be that the entire release
is to the atmosphere.

7.13. If multiple release routes for a single accident scenario are considered (which
might be necessary if the same groups of people could be significantly affected as

a result of each release route), then the results from each separate modelling study should be combined to determine the dose to the representative person.

IDENTIFICATION OF THE REPRESENTATIVE PERSON AND REFERENCE ANIMALS AND PLANTS

7.14. Reference [19] provides guidance on selecting representative persons and determining their habits (e.g. location, time spent outdoors, breathing rate, food consumption rate) so that their overall radiological exposure can be determined. The selection should be reasonable, sustainable and homogeneous. In other words, the selection should not be overly conservative by combining a set of extreme habits or describing one-off events but should instead consider situations likely to continue over the lifetime of the nuclear installation and applying to more than a single individual.

7.15. The representative person may be different for operational states and accident conditions at the same site or installation. Possible representative persons for an accidental release from a nuclear installation could be as follows:

(a) An adult spending most of the time in a location leading to a higher likelihood of exposure, for example in the prevailing wind direction close to the site boundary/exclusion area boundary or close to sediment in surface waters that have accumulated radioactivity;
(b) An adult consuming a high rate of local food and living in the nearest population centre;
(c) A child or infant consuming supermarket food or non-locally produced food and living in the nearest population centre;
(d) An adult not resident in the nearest population centre, but who spends a significant fraction of the day working outdoors close to the release point (e.g. in an agricultural occupation on an adjacent site);
(e) A pregnant woman's fetus.

7.16. Several conservative assumptions should not be combined in selecting the representative person. For example, an infant living in the most exposed location and consuming only local food should not be selected.

7.17. To assess the total risk from a nuclear installation, the risks to the representative person and/or the societal consequences from each accident scenario, and the respective scenario frequency, should be considered.

7.18. For an atmospheric release, the radiological consequences are strongly dependent on the meteorological conditions at the time of the release. For example, the number of people exposed and the level of exposure for individuals are very different for various wind directions and other meteorological conditions. Different approaches may be applied, from a deterministic assessment using reasonably bounding meteorological conditions to statistical methods accounting for the many variabilities (e.g. meteorological conditions, exposure scenarios). One such option is a Level 3 probabilistic safety assessment[10], but the choice of an adequate option depends on the risk criteria to be met. With a lower risk installation, a proportional approach with simpler modelling conditions (e.g. a single set of bounding meteorological conditions) may be acceptable.

7.19. Statistical methods usually sample from a data set of hourly historic data measured or calculated for the site, as described in Section 4. Since the accident scenario considered in the assessment might extend over many hours or days, the meteorological data for each hour of the release or for discrete phases should be used; this is termed a meteorological sequence. Sufficient meteorological sequences should be sampled to represent the range of meteorological conditions that a given site might experience. For random meteorological sampling, the relative likelihoods of each sequence are unity; if other sampling schemes are used to obtain more examples of particular types of weather (e.g. heavy rainfall), then the relative likelihoods should be adjusted accordingly. This procedure should be repeated for every accident scenario considered, whether it is a part of a deterministic assessment or part of a probabilistic assessment. In this latter case, the conditional risks for each scenario are multiplied by their respective frequencies (derived from the Level 1 and/or Level 2 probabilistic safety assessment [18, 23]) to give the risk for each scenario. An analogous procedure could be followed for aquatic releases, if deemed necessary.

7.20. Computer codes used for statistical methods may combine several aspects such as atmospheric dispersion modelling and dose and risk calculations using meteorological data, demographic data and food production data for the nuclear installation site and surrounding area. When using such codes or specific models developed for different environments, care should be taken to ensure they are applicable to the environment being assessed. For example, such codes might use food chain models developed for agricultural practices in temperate climates that might not be suitable for use in tropical climates.

---

[10] See footnote 6 on p. 39.

7.21. Where exposure of animals and plants is considered, this is usually only for normal operation. Care should be taken to protect the more highly exposed population groups of a species rather than individual members of species within those groups.

7.22. Animals and plants might occupy different habitats to those occupied by humans and might be exposed to radionuclides that accumulate over time in those habitats. A region around the discharge locations of the nuclear installation (e.g. with an area of around 100–400 km^2) should be considered, in which population groups of animal and plant species that might be exposed are identified. A set of reference animals and plants is defined in Ref. [28]. A generic methodology for assessing exposures of flora and fauna is provided in annex I of GSG-10 [13].

ASSESSMENT OF DOSE AND/OR RISK TO THE REPRESENTATIVE PERSON AND TO REFERENCE ANIMALS AND PLANTS

7.23. The radiological impact to the public from normal operation should be assessed on the basis of the individual effective dose to the representative person. Whether a quantitative assessment of the radiological impact on fauna and flora from normal operation is also needed is a matter for individual States. With regard to this, para. 1.33 of GSR Part 3 [4] states:

> "Trends also indicate the need to be able to demonstrate (rather than to assume) that the environment is being protected against effects of industrial pollutants, including radionuclides, in a wider range of environmental situations, irrespective of any human connection. This is usually accomplished by means of a prospective environmental assessment to identify impacts on the environment, to define the appropriate criteria for protection of the environment, to assess the impacts and to compare the expected results of the available options for protection. Methods and criteria for such assessments are being developed and will continue to evolve."

7.24. The components of the radiological impact on the public from accidental releases is shown in fig. 3 of GSG-10 [13].

7.25. The framework for radiation protection of members of the public and protection of the environment in planned exposure situations, emergency exposure situations and existing exposure situations and its application are provided in IAEA Safety Standards Series No. GSG-8, Radiation Protection of the Public and the Environment [29]. Recommendations on the application of the principles of justification, optimization of protection and dose limits, where appropriate, are provided in GSG-8 [29].

DETERMINATION OF THE ACCEPTABILITY OF RADIOLOGICAL IMPACTS

7.26. For operational states (i.e. normal operation and anticipated operational occurrences), recommendations on setting dose limits and constraints are provided in GSG-9 [20].

7.27. For potential exposures, para. 3.15(e) of GSR Part 3 [4] requires that the likelihood and magnitude of such exposures, their likely consequences and the number of individuals who may be affected be assessed, while para. 3.120 of GSR Part 3 [4] requires that constraints on dose and constraints on risk be established but no criteria are specified. It is the responsibility of the government or the national regulatory body to either approve or establish dose constraints. International guidance on determining the acceptability of impacts is provided in Refs [15, 30].

7.28. Impacts on neighbouring States from both operational states and accident conditions should be considered. Paragraph 3.124 of GSR Part 3 [4] states:

"When a source within a practice could cause public exposure outside the territory or other area under the jurisdiction or control of the State in which the source is located, the government or the regulatory body:

(a)   Shall ensure that the assessment for radiological impacts includes those impacts outside the territory or other area under the jurisdiction or control of the State;

....

(c)   Shall arrange with the affected State the means for the exchange of information and consultations, as appropriate."

7.29. For nuclear power plants, since it is difficult to exclude the possibility of any public exposure in neighbouring States, a transboundary assessment should be performed. For nuclear installations other than nuclear power plants, a transboundary assessment should be conducted if a facility has the potential to affect an area across borders.

7.30. When considering transboundary impacts, the criteria used for the assessment of the level of protection for operational states or for the consideration of potential exposures in other States should be in line with the criteria set out in GSG-10 [13] and should be the same as those used for the State in which the installation is located.

7.31. GSR Part 3 [4] establishes general requirements on environmental protection from radiation but does not have specific requirements on the assessment of exposure (and hence the level of protection) for flora and fauna. GSR Part 3 [4] notes the importance of assessing environmental impacts, while allowing decision flexibility commensurate with radiation risks associated with the facility or activity. If regulations require a flora and fauna exposure assessment, typical environmental monitoring programmes for public protection, as detailed in GSG-10 [13] and GSG-19 [14], are generally sufficient to validate the exposure assessment. The IAEA follows (to the extent applicable) ICRP recommendations in publications 103, 108, 114, 124 and 136 [28, 31–34], which use 12 reference animals and plants to assess radiological impacts on non-human species. This approach, aligned with human protection principles, includes derived consideration reference levels and has been incorporated into GSG-10 [13] for planned exposure situations.

7.32. As part of the application for a licence for a new nuclear installation project, the applicant prepares, and the regulatory body reviews, an environmental management plan, which is a comprehensive document that identifies, among other things, the actions to be taken (including any mitigation measures that are included in the environmental impact assessment report and licensing conditions imposed by the regulatory body), responsibilities, reporting and processes for implementing corrective actions if needed. The purpose of the environmental management plan is to ensure that potential project interactions with the environment are considered during the site preparation, construction, operation and decommissioning stages of the nuclear installation, to minimize or prevent potential negative impacts and to enhance the positive impacts. Some States might not require such a combined document but instead might require individual plans for specific issues. The environmental monitoring programme, which is presented in Section 8 of this Safety Guide, should be a part of the environmental management plan. That monitoring programme may need adjustment during the decommissioning process owing to the changing nature of activities at the site.

8. MONITORING OF RADIOACTIVITY IN THE ENVIRONMENT AROUND A NUCLEAR INSTALLATION

8.1. Requirement 28 of SSR-1 [1] states that **"All natural and human induced external hazards and site conditions that are relevant to the licensing and safe operation of the nuclear installation shall be monitored over the lifetime of the nuclear installation."**

ENVIRONMENTAL MONITORING PROGRAMME DURING THE SITE CHARACTERIZATION AND PRE-OPERATIONAL STAGES OF A NUCLEAR INSTALLATION

8.2. Recommendations on the measurement of the background radioactivity as part of the establishment of baseline environmental conditions at the site and in the vicinity of the site for a nuclear installation are provided in Section 3. Sampling programmes should be developed for measuring background radiation and other parameters used to estimate radiation doses due to direct radiation from radioactive material inside the installation and from radioactive material released from the installation to the air, surface water, groundwater and the ground surface during the site characterization and pre-operational stages. These sampling programmes can serve as the basis for the monitoring programmes that will be established during the operation of the proposed nuclear installation. The locations and the media chosen for measuring the radioactivity and other parameters should be those that are likely to be relevant to the exposure of representative persons when the nuclear installation starts to operate.

8.3. The environmental monitoring programme should commence well before the start of construction of the installation and sufficiently before operation to be able to identify any trends in the background levels of radioactivity. For example, if the levels of a particular radionuclide are falling prior to the start of operation, then they would be expected to continue to fall in the absence of any new releases from the installation.

8.4. The environmental monitoring programme should continue for the lifetime of the installation.

Monitoring of radioactivity in the atmosphere

8.5. Monitoring stations should be set up at several key locations on the site, on the site perimeter and away from the site to measure external radiation and the radioactivity in materials suspended in the air. These stations should initially be used to establish the background radiation and meteorological conditions in the area. Their operation should continue after the installation starts operating, to determine changes in radioactivity in the air due to the operation of the installation and to record any changes in the meteorological conditions at the site and in its vicinity.

Monitoring of radioactivity in surface water and groundwater

8.6. A monitoring programme should be established for both surface water and groundwater. The purposes of such a programme during the site characterization and pre-operational stages are to establish the baseline conditions and to determine whether there are any trends that result in changing the characteristics of the region before the commencement of operation of the nuclear installation. Recommendations on the selection of sampling locations for monitoring of surface water and groundwater are provided in GSG-19 [14].

8.7. All surface water and groundwater in the vicinity of the site should be sampled regularly. For surface water bodies, sediments should be sampled as well as the water itself.

8.8. Groundwater should be monitored by means of samples taken from boreholes and wells. The samples can also be taken from groundwater reaching the surface in springs or in natural depressions. Boreholes and wells should be kept in an operable state.

8.9. The monitoring programme for groundwater should be initiated at least two years before the start of construction of the installation. The site area should be monitored before the foundation work is begun in order to verify possible changes in the groundwater regime. Monitoring should continue after construction has finished.

Monitoring of radioactivity in soils and biota

8.10. Soil samples should be taken and analysed to determine the baseline radionuclide concentrations in the soils before the start of operation of the nuclear installation (see Section 3). The sampling locations should include areas on the site, in particular those that will not be covered by buildings or paved over by roads and parking lots, as well as areas off the site that are residential, industrial, commercial, agricultural, recreational or for wildlife.

8.11. Samples should also be taken and analysed to determine the baseline concentrations of radionuclides in biota in the vicinity of the nuclear installation. Biota samples should include the following:

(a)   Foodstuffs grown (e.g. vegetables, fruits, grains) and animals reared in the region for human consumption;
(b)   Biota consumed by domesticated and wild animals (e.g. grass, leaves on bushes, shrubs, low tree branches);
(c)   Wild animals in the region (e.g. birds, rabbits).

Guidelines on soil and vegetation sampling for radiological monitoring are given in Ref. [35].

Monitoring of population data and other parameters

8.12. The monitoring programmes started during the pre-operational stages of a nuclear installation and continued during operation and decommissioning should focus on those radionuclides that are important contributors to the total dose of the representative person. In addition, those parameters that are identified as important to this dose calculation through modelling studies and sensitivity analyses should be sampled more frequently and in more locations than other parameters to which the results are less sensitive. The distance of the sampling locations from the nuclear installation should be determined by the results of the pathway analyses. If the results indicate that an individual could receive a substantial dose through a pathway at some distance, the environmental samples should be extended to that distance. These distances vary for different types of nuclear installations depending on the source terms and site environmental factors. The control locations (see para. 3.36) that are outside the region of influence of the nuclear installation should continue to be sampled regardless of their distance from the installation.

8.13. Arrangements for emergency preparedness should be considered carefully for any conceivable emergency when implementing the monitoring programmes during the pre-operational stage (see GSG-19 [14]). The basic intervention levels should be understood by all responsible persons and organizations, and operational intervention levels should be established on a site specific basis. The operational intervention levels should refer to parameters that can be easily measured (e.g. dose rate in air, deposition density of radionuclides) so that an interpretation can be made rapidly if intervention is needed.

MONITORING PROGRAMME DURING THE OPERATION OF A NUCLEAR INSTALLATION

8.14. Requirements for monitoring during operation of a facility are established in Requirements 14, 20 and 32 of GSR Part 3 [4]. GSG-19 [14] provides supporting recommendations and Ref. [36] provides detailed descriptions of the monitoring programmes employed during the operation of a nuclear installation, including information about the objectives, conduct and use of monitoring both during operational states and in accident conditions.

8.15. During the operation of a nuclear installation, monitoring programmes should be used for the following purposes:

— To verify compliance with regulatory limits such as dose constraints;
— To confirm that the levels of radionuclides in the environment are consistent with the discharges reported by the operating organization;
— To confirm the results of the impact study;
— To check the predictions of environmental models;
— To provide a warning of unusual or unforeseen conditions.

In an emergency, additional monitoring activities should be established.

Environmental monitoring during normal operation

8.16. The environmental monitoring programme established during the site characterization and pre-operational stages of a nuclear installation should be continued during the operation of the installation. Samples of environmental media should be taken and analysed on a schedule that depends on the half-lives of the radionuclides that could potentially be discharged from the installation, their discharge route and also corresponding to the objective of the analysis to be made. However, the frequency and the number of samples taken during the early stages of operation of the installation should be relatively high to confirm the predictions made by modelling conducted during the site characterization and pre-operational stages. As experience is gained, the scale of routine monitoring could be reduced and the locations amended to reflect actual discharge patterns identified during monitoring activities.

8.17. Environmental monitoring in the context of this Safety Guide refers to the measurement of external dose rates in the environment and radionuclide activity concentrations in air, water, soil, sediments, vegetation, foodstuffs and the bodies of animals. A key feature in designing environmental monitoring programmes for major sources is the identification of potentially critical radionuclides, pathways and groups. Once identified and assessed, it is possible to select those radionuclides and pathways that make the biggest contribution to individual doses so that the monitoring programmes can be focused accordingly.

Environmental monitoring during a nuclear or radiological emergency

8.18. Environmental monitoring takes on special significance during an emergency because it often provides important information about the severity of impacts from the accident. GSR Part 7 [5] requires prompt monitoring and assessment of areas that could be or are known to have been contaminated during a nuclear or radiological emergency. GSG-19 [14] provides recommendations on environmental monitoring in an emergency.

Source monitoring

8.19. Source monitoring is the monitoring of a particular source of radiation or discharges of radionuclides to the environment arising from a nuclear installation. Source monitoring programmes are usually designed to measure dose rates at the source and/or the discharge rates of radionuclides, which may be in the form of gases, aerosols or liquids. The results from source monitoring can be used to verify compliance with the authorized limits on discharges and/or as a basis for estimating environmental radiation levels and activity concentrations in environmental media using predictive modelling. The results of source monitoring can also provide an early warning of any deviations from normal operation.

8.20. In accordance with GSG-19 [14], there should be coordination between the source monitoring programme and the environmental monitoring programme. In the case of discharges where the activity concentrations in environmental monitoring are below detection limits, the dose calculations can be based on source monitoring data and appropriate modelling.

8.21. Further recommendations on source monitoring are provided in GSG-19 [14].

Individual monitoring

8.22. Individual monitoring relates to measurements taken directly on people. It includes measurements of external doses with personal dosimeters carried by individuals and/or measurements of the quantities of radioactive substances in the body or in excreta, and the interpretation of such measurements in terms of individual dose. Workers who are exposed to radiation at varying levels in different parts of the nuclear installation site are routinely monitored.[11] Members of the public who visit the site may also be monitored. However, members of the public off the site are not normally monitored individually.

8.23. Paragraph 5.52 of GSR Part 7 [5] states:

> "The operating organization and response organizations shall ensure that arrangements are in place for the protection of emergency workers and protection of helpers in an emergency for the range of anticipated hazardous conditions in which they might have to perform response functions. These arrangements, as a minimum, shall include:
>
> ....
>
> (c) Managing, controlling and recording the doses received".

8.24. Requirement 12 of GSR Part 7 [5] states:

> **"The government shall ensure that arrangements are in place for the provision of appropriate medical screening and triage, medical treatment and longer term medical actions for those people who could be affected in a nuclear or radiological emergency."**

8.25. Recommendations on individual monitoring in an emergency exposure situation are provided in GSG-19 [14].

---

[11] Recommendations on the monitoring of workers and the workplace are provided in IAEA Safety Standards Series No. GSG-7, Occupational Radiation Protection [37], and further information is provided in Ref. [38]).

Other monitoring situations

8.26. Paragraph 3.137 of GSR Part 3 [4] states:

"Registrants and licensees shall, as appropriate:

(a) Establish and implement monitoring programmes to ensure that public exposure due to sources under their responsibility is adequately assessed and that the assessment is sufficient to verify and demonstrate compliance with the authorization. These programmes shall include monitoring of the following, as appropriate:
    (i) External exposure due to such sources;
    (ii) Discharges;
    (iii) Radioactivity in the environment;
    (iv) Other parameters important for the assessment of public exposure."

8.27. For a nuclear installation, the 'other parameters' mentioned in the quotation in para. 8.26 should include the following:

(a) Population distribution (resident and transient) and characteristics (e.g. age, gender);
(b) Population habits (e.g. food consumption rates, proportion of time that people spend indoors and outdoors);
(c) Agricultural activity in the region (e.g. types and quantities of food grown);
(d) Types and numbers of animals raised for food in the region;
(e) Wildlife in the region;
(f) The proportion of locally grown food that is consumed in the region compared with that exported outside the region.

8.28. If the operating organization of the nuclear installation included any mitigation measures in its assessment of the radiological impacts when applying for authorization, or if the regulatory body granting authorization imposed any conditions, the parameters needed to verify and document those measures or conditions should also be addressed in the monitoring programme.

8.29. Certain non-radiological impacts of a nuclear installation are usually included in the environmental impact assessment as part of the authorization process. These include the impact on people and the environment from releases of hazardous chemicals and heated water, the impact from the construction of the installation, the impact on places of societal significance (e.g. historical monuments, cultural places), the impact on endangered species and the impact on the landscape, as well as other societal and economic factors. Such impacts are normally considered by the regulatory body, taking into account regulatory requirements. It may be cost effective and beneficial to coordinate any monitoring activities that are necessary as part of the non-radiological impact assessment with the activities undertaken in the radiological monitoring programme.

MONITORING PROGRAMME FOLLOWING PERMANENT SHUTDOWN

8.30. Paragraph 9.3 of IAEA Safety Standards Series No. GSR Part 6, Decommissioning of Facilities [39], states:

"If the approved decommissioning end state is release from regulatory control with restrictions on the future use of the remaining structures, appropriate controls and programmes for monitoring and surveillance shall be established and maintained for the optimization of protection and safety, and protection of the environment."

8.31. With regard to nuclear fuel cycle facilities, para. 5.13 of SSR-4 [3] states:

"The operating organization shall establish a programme of monitoring throughout the lifetime of the facility to evaluate natural and human-made changes in the area, including changes in demographics. The programme of monitoring shall be in place no later than the start of construction and shall continue through to decommissioning until termination of the authorization."

8.32. When a nuclear installation ceases to operate and before the start of decommissioning, and in accordance with the decommissioning plan, the monitoring programme should be re-evaluated and modified as appropriate for the decommissioning stage and for the stage from the end of decommissioning to release from regulatory control. Further recommendations are provided in GSG-19 [14] and IAEA Safety Standards Series No. SSG-47, Decommissioning of Nuclear Power Plants, Research Reactors and Other Nuclear Fuel Cycle Facilities [40], and additional information is provided in Ref. [41].

9. CONSIDERATION OF THE FEASIBILITY OF PLANNING EFFECTIVE EMERGENCY RESPONSE ACTIONS

9.1. Requirement 13 of SSR-1 [1] states:

"The feasibility of planning effective emergency response actions on the site and in the external zone shall be evaluated, with account taken of the characteristics of the site and the external zone as well as any external events that could hinder the establishment of complete emergency arrangements prior to operation."

9.2. Any adverse conditions surrounding the site that could hinder off-site emergency response actions, such as the sheltering or evacuation of the population in the region or the access of external services needed to deal with an emergency, should be identified and evaluated (e.g. by performing a transport analysis; see para. 9.9) and it should be confirmed that planning effective emergency response actions remains feasible.

9.3. The area to consider should be large enough to encompass any possible future emergency planning zone or distance (see para. 5.38 of GSR Part 7 [5] and appendix II of IAEA Safety Standards Series No. GS-G-2.1, Arrangements for Preparedness for a Nuclear or Radiological Emergency [42]).

9.4. Geographical features of the landscape that might make off-site protective actions difficult to implement include physical barriers such as rivers or mountains that lack infrastructure for facilitating a response. Feeder roads (e.g. frontage roads, spurs, secondary roads) that offer single points of access are not barriers to response as they lead to major evacuation routes. If evacuation is likely to be necessary, then, as a high level principle, there should be at least two major evacuation routes in different directions to offer various itinerary options for the implementation of precautionary urgent protective actions or urgent protective actions that involve road transportation during a nuclear or radiological emergency. If this is not possible owing to geographical features, administrative restrictions or other reasons, the site should be considered unsuitable for a nuclear installation unless measures are identified that would, when implemented, mitigate or eliminate the barriers to response. Examples of unsuitable sites are provided in Figs 2 and 3, while an example of a suitable site is provided in Fig. 4.

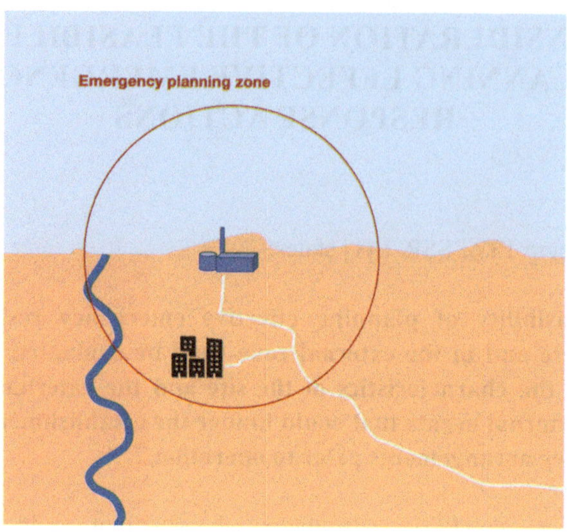

FIG. 2. Example of a site with a physical barrier (river) preventing construction of an alternative evacuation route in another direction. The site is unsuitable unless a bridge is constructed or an alternative evacuation route is made available.

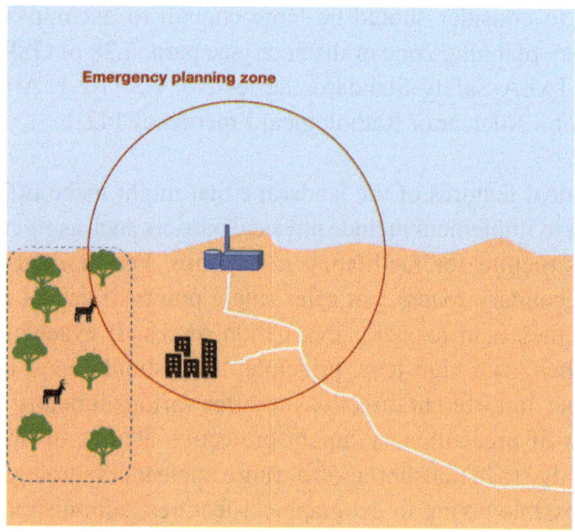

FIG. 3. Example of a site with an administrative barrier (national park or special area) preventing construction of an alternative evacuation route in another direction. The site is unsuitable.

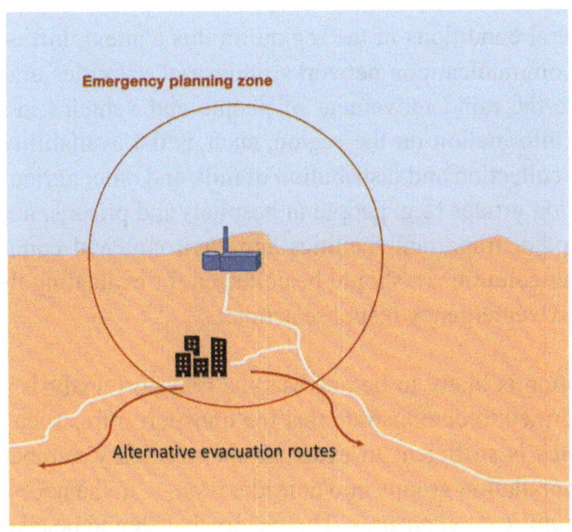

FIG. 4. Example of a suitable site. Two major evacuation routes (with sufficient capacity) in different directions exist or can be constructed.

9.5. In preparedness for a nuclear or radiological emergency, effective arrangements for sheltering should include the identification in advance of large and closed buildings that can be easily accessed by the public. These buildings, particularly in warmer climates, should be assessed for their effectiveness as shelters, considering that contamination can pass through even small openings.

9.6. As indicated in appendix V to GS-G-2.1 [42], iodine thyroid blocking tablets should be provided to the local population before or shortly after an intake of radioiodine. If iodine thyroid blocking tablets have not been pre-distributed and an order to administer the tablets is issued during a nuclear emergency, an absence of infrastructure (e.g. insufficient access options by road) might hinder access to iodine thyroid blocking tablets for the local population. However, if the site is suitable for evacuation, it can be assumed that the infrastructure also supports the distribution of tablets.

9.7. Restrictions may need to be placed on the consumption of food and water that can become contaminated from a release. Although these might need to be established in the hours or days following a release, such restrictions have no implications for site suitability.

9.8. The feasibility of planning effective emergency response actions for the nuclear installation site should be evaluated on the basis of the specific natural and infrastructural conditions in the region. In this context, infrastructure means transport and communications networks, industrial activities and anything that might influence the rapid movement of people and vehicles in the vicinity of the site. Other information on the region, such as the availability of sheltering, systems for the collection and distribution of milk and other agricultural products, special population groups (e.g. people in hospitals and prisons, nomadic groups), industrial and other important facilities and environmental conditions (e.g. the range of weather conditions), should be collected for evaluating the feasibility of planning effective emergency response actions.

9.9. If evacuation is likely to be necessary, a transport analysis (e.g. road, rail) should be performed to demonstrate that the transport infrastructure for multiple evacuation routes is sufficient to evacuate the necessary number of people — taking special population groups into consideration — in the necessary amount of time to avoid significant exposures. The factors that should be taken into account for the transport analysis of each alternative route include the following:

(a)   Number of people to evacuate;
(b)   Available vehicles;
(c)   Transport needs and arrangements for any special population groups;
(d)   Time to alert people and to prepare to evacuate, taking into consideration any special population groups;
(e)   Typical traffic volumes;
(f)   Typical road capacity;
(g)   Traffic bottlenecks such as bridges.

9.10. When planning for an evacuation, the transport infrastructure does not need to be present at the time of the site evaluation, but it should be practicable to improve the infrastructure so that the off-site emergency plan can be made feasible before operation.

9.11. Many site related aspects should be considered in assessing the feasibility of planning effective emergency response actions. The most important aspects are:

(a)   Population density and distribution in the region;
(b)   Distance of the site from population centres;
(c)   Special population groups;
(d)   Particular geographical features, such as islands, mountains and rivers;
(e)   Characteristics of local transport and communications networks;
(f)   Industrial facilities where potentially hazardous activities are conducted;

(g)   Agricultural activities that are sensitive to possible discharges of radionuclides;

(h)   Possible combinations of hazards from external events (e.g. earthquake with flooding).

9.12. The presence of large populations in the region or the proximity of a city to the nuclear installation should be carefully considered in the hazard assessment to develop effective off-site emergency arrangements. The specific circumstances of any special population groups should be recognized and taken into account.

9.13. External events might have consequences that limit the effectiveness of the response to an emergency at a nuclear installation. For example, an external event might result in a problem with the infrastructure or in damage to sheltering facilities. In order to ensure that the population in the region can be sheltered and evacuated effectively, consideration should be given to the provision of backup facilities and alternative routes. External events such as earthquakes might damage key transport infrastructure such as bridges; this may need to be considered when assessing alternative transport scenarios.

9.14. If it is determined that no effective response actions can be established, then the proposed site should be considered unsuitable.

9.15. The conditions assessed for the approval of the site and design may change over time. The site characteristics considered in the off-site emergency plan, such as infrastructural developments, should therefore be reviewed periodically during the operational stage of the nuclear installation.

10. APPLICATION OF A GRADED APPROACH TO RADIOLOGICAL ENVIRONMENTAL IMPACT ASSESSMENT FOR NUCLEAR INSTALLATIONS

10.1. A graded approach[12] should be applied to radiological environmental impact assessment for nuclear installations, on the basis of the following:

(a) The stage in the lifetime of the nuclear installation;
(b) The level of complexity of the site and the radionuclide transfer mechanisms;
(c) The hazard category of the installation (see para. 10.5).

10.2. The advantages of applying a graded approach include the following:

(a) It allows conservative calculations to be made, reducing the time and cost of data acquisition.
(b) It avoids the construction of models more complicated than needed to make sufficiently useful and accurate predictions.
(c) It provides a guideline for selecting a suitable level of complexity for reporting appropriate to the specific stage of site evaluation.

10.3. Radiological environmental impact assessment is required to be performed using site specific data and site specific design parameters (if the technology is known; otherwise a plant parameter envelope can be used in the interim and updated at a later stage) during site characterization (see para. 4.44 of SSR-1 [1]). During further stages of site evaluation over the lifetime of the nuclear installation, it may be necessary to update the assessments, depending on the availability of data or results of the monitoring programmes. The radiological environmental impact during construction and operation should be confirmed by the environmental monitoring programme. If the impacts deviate from those expected, a process should be initiated to determine the cause of the discrepancies and implement remedial measures if necessary.

10.4. The level of complexity incorporated in a model should be commensurate with the purpose, hazard category (see para. 10.5) and stage of site evaluation

---

[12] The use of a graded approach is intended to ensure that the necessary levels of analysis, documentation and actions are commensurate with, for example, the magnitudes of any radiological hazards and non-radiological hazards, the nature and the particular characteristics of a facility, and the stage in the lifetime of a facility [6].

for the installation. The radionuclide transfer mechanisms included should be based on the complexity of the system to achieve acceptable accuracy. The system may be simplified to make a first approximation, or complexities may be incorporated implicitly if their effect on transfer is deemed less relevant. Simplifications might decrease the accuracy of the model and do not necessarily mean that the results are conservative. The modelling may be in one, two or three dimensions, assuming steady state or transient flow conditions. One and two dimensional modelling assuming steady state conditions is much easier than three dimensional modelling and should be performed for low hazard installations and for exploratory purposes for high hazard installations to determine the level of complexity that should be applied in the subsequent stages. Analytical models may be used for low to intermediate hazard installations. For intermediate and high hazard installations, a stepwise strategy may be followed. Modelling with some simplifications of the site characteristics and use of the most conservative transfer mechanisms may precede more detailed modelling that takes into account site complexities (e.g. stratification in lakes, uniform or non-uniform geometry in rivers, heterogeneity and anisotropy in groundwater systems, complex surface topography for dispersion in the atmosphere) and known mechanisms (e.g. dispersion, sorption and first-order reaction in groundwater systems, sedimentation or resuspension in rivers and lakes).

10.5. The radiological environmental impact assessment for nuclear installations should be commensurate with the radiological hazards and with the hazards due to other materials present on the site. In general, the criteria for categorization should be based on the radiological consequences of the release of radioactivity from the installation, ranging from very low radiological consequences to potentially severe radiological consequences. As an alternative, the categorization may range from radiological consequences within the installation itself, to radiological consequences within the site boundary, to radiological consequences to the public and the environment outside the site. Three or more categories of nuclear installation may be defined (e.g. those presenting high, medium or low hazards) on the basis of national practice. The analysis process may be performed iteratively, where complexity is sequentially added until no more complexity in the analysis is necessary. For the application of a graded approach, the hazard categorization of a nuclear installation can be based on the same characteristics as listed in para. 9.5 of SSG-9 (Rev. 1) [8], as follows:

(a) The amount, type and status of the radioactive inventory at the site (e.g. whether solid, liquid and/or gaseous, whether the radioactive material is being processed or only stored);

(b)   The intrinsic hazard associated with the physical processes (e.g. nuclear chain reactions) and chemical processes (e.g. for fuel processing purposes) that take place at the installation;

(c)   The thermal power of the nuclear installation, if applicable;

(d)   The configuration of the installation for different kinds of activity (depending on the design and accident management, there might be considerable time before a major release of radioactive material is initiated);

(e)   The distribution of radioactive sources in the installation (for research reactors, most of the radioactive inventory is in the reactor core and the fuel storage pool, whereas for fuel processing and storage facilities it might be distributed throughout the installation);

(f)   The changing nature of the configuration and layout of installations designed for experiments (such activities have an associated intrinsic unpredictability);

(g)   The need for active safety systems and/or operator actions for the prevention of accidents and for mitigation of the consequences of accidents, and the characteristics of engineered safety features for the prevention of accidents and for mitigation of the consequences of accidents (e.g. the containment and containment systems);

(h)   The characteristics of the structures of the nuclear installations and the means of confinement of radioactive material;

(i)   The characteristics of the processes or of the engineering features that might show a cliff edge effect[13] in the event of an accident;

(j)   The characteristics of the site that are relevant to the consequences of the transfer of radioactive material to the atmosphere and the hydrosphere (e.g. size and demographics of the region);

(k)   The potential for on-site and off-site contamination.

10.6. The application of a graded approach may allow certain simplifications and a less detailed approach in the following areas:

(a)   Source term (e.g. radionuclide quantity, activity, mass and/or volume, form, chemical and physical composition, geometry, height of release, potential for release, release start time and time profile of the release, novelty of design or activity);

---

[13] In a nuclear power plant or nuclear fuel cycle facility, a cliff edge effect is an instance of severely abnormal facility behaviour caused by an abrupt transition from one facility status to another following a small deviation in a facility parameter, and thus a sudden large variation in facility conditions in response to a small variation in an input [6].

(b)   Complexity of environmental characteristics of the site and its region: characteristics of the site and its region relating to dispersion of radionuclides in the environment (e.g. hydrogeology, hydrology, meteorology, morphology, biophysical characteristics), presence and characteristics of receptors (e.g. demography, population habits and living conditions, flora and fauna, exposure pathways), land use and other activities (e.g. agriculture, food processing, other industries) and characteristics of other installations in the vicinity;
(c)   Dimensionality of model (i.e. one, two or three dimensional);
(d)   Steady state and transient transfer mode;
(e)   Type of model (i.e. analytical, numerical or statistical);
(f)   Source of information;
(g)   Transfer phenomena.

10.7. An example of the application of a graded approach to the analysis of radionuclide dispersion in groundwater is given in the Appendix.

11. APPLICATION OF THE MANAGEMENT SYSTEM TO THE INVESTIGATION OF SITE CHARACTERISTICS AND EVALUATION OF RADIATION RISKS FROM A NUCLEAR INSTALLATION

11.1. A management system is required to be established, applied and sustained by senior management (see Requirement 3 of IAEA Safety Standards Series No. GSR Part 2, Leadership and Management for Safety [43]). This applies to all facilities and activities and should be implemented for the activities that are performed for the investigation of site characteristics and evaluation of radiation risks to the public and the environment in site evaluation for nuclear installations.

PROJECT MANAGEMENT APPROACH

11.2. A project work plan for the investigation of site characteristics and evaluation of radiation risks from a nuclear installation should be established that, at a minimum, addresses the following topics:

(a) The objectives and scope of the project;
(b) Applicable regulations and standards;
(c) Organization of the roles and responsibilities for management of the project;
(d) Work breakdown, processes, tasks, schedule and milestones;
(e) Interfaces among the different types of tasks (e.g. data collection tasks, analysis tasks) and disciplines involved, especially the various specialists needed for the different aspects of the investigation of site characteristics and evaluation of radiation risks to the public and the environment with all necessary inputs and outputs;
(f) Project deliverables and reporting.

11.3. The project scope should identify all aspects of the investigation of site characteristics and evaluation of radiation risks that are relevant to the impact of the nuclear installation on the public and the environment and that are investigated within the framework of the project.

11.4. The project work plan should include a description of all requirements that are relevant for the project, including applicable regulatory requirements in relation to the investigation of site characteristics and evaluation of radiation risks to the public and the environment. The applicability of the set of regulatory requirements should be reviewed by the regulatory body prior to conducting the project activities.

11.5. All approaches and methodologies that reference lower tier legislation (e.g. regulatory guidance documents, industry codes and standards) should be clearly identified and described. The details of the approaches and methodologies to be used should be clearly stated in the project work plan.

11.6. At a minimum, the following generic processes should be included in the management system to ensure quality of the project:

(a) Document control;
(b) Control of products;
(c) Controls for measuring and testing equipment;
(d) Control of records;

(e)   Control of analyses;
(f)   Purchasing (procurement);
(g)   Validation and verification of software;
(h)   Validity and quality of data;
(i)   Audits (self-assessment, independent assessments and review);
(j)   Control of non-conformances;
(k)   Corrective actions;
(l)   Preventive actions.

Processes covering field investigations, laboratory testing, data collection and the analysis and evaluation of observed data, as well as communication processes for interaction among the experts involved in the project, should also be applied.

11.7. The project work plan should ensure that there is adequate provision, in terms of resources and in the schedule, for collecting and analysing new data that might be important for the conduct of the investigation of site characteristics and evaluation of radiation risks to the public and the environment.

11.8. To make the investigation of site characteristics and evaluation of radiation risks to the public and the environment traceable and transparent to users (e.g. peer reviewers, the operating organization, the regulatory body), the documentation for the analysis should provide a description of all elements of the process and include the following information:

(a)   A description of the study participants and their roles;
(b)   Background material that includes documentation on the data collected and analysed;
(c)   A description of the computer software used, and the input and output files;
(d)   Reference documents;
(e)   All documents supporting the treatment of uncertainties and related discussions;
(f)   Results of intermediate calculations and sensitivity studies.

This documentation should be maintained in an accessible, usable and auditable form by the operating organization.

11.9. The documentation should include references to all sources of information used in the investigation of site characteristics and evaluation of radiation risks to the public and the environment, including information on the sources of important citations that might be difficult to trace.

ENGINEERING USES AND OUTPUT SPECIFICATION

11.10.   The investigation of site characteristics and evaluation of radiation risks to the public and the environment should be conducted in order to develop the site evaluation report and environmental impact assessment report. From the beginning, the work plan for the investigation of site characteristics and evaluation of radiation risks to the public and the environment should identify the intended engineering uses and objectives and should incorporate an output specification that describes all the results necessary for the intended engineering uses and objectives of the project.

DOCUMENTATION OF THE INVESTIGATION OF SITE CHARACTERISTICS AND EVALUATION OF RADIATION RISKS TO THE PUBLIC AND THE ENVIRONMENT

11.11.   The project for the investigation of site characteristics and evaluation of radiation risks to the public and the environment should be well documented, with a clearly defined scope and objectives. The conceptual models used for numerical modelling should be described in detail. The code selected and the reasons for its selection should be described. The steps of model construction should be documented, including grid construction, assignment of parameters, boundary conditions, steady state and transient calibration, and sensitivity analysis, if applicable. Simulation runs should be documented. The scenarios should be well described, and the results should be discussed, taking into consideration the uncertainties. An electronic copy of a ready-to-run model should be provided as an appendix to the documentation. The electronic copy should include the inputs and outputs of each run and a description of the version of software used and the operating system it was used on.

INDEPENDENT PEER REVIEW OF THE INVESTIGATION OF SITE CHARACTERISTICS AND EVALUATION OF RADIATION RISKS TO THE PUBLIC AND THE ENVIRONMENT

11.12.   An independent peer review should be conducted to provide assurance of the following:

(a)   That a proper process has been duly followed in conducting the investigation of site characteristics and evaluation of radiation risks to the public and the environment;

(b)   That the analysis addresses the uncertainties involved;

(c)   That the documentation is complete and traceable.

11.13.   The independent peer review team should possess the multidisciplinary expertise needed to address all technical and process related aspects of the investigation of site characteristics and evaluation of radiation risks to the public and the environment. The team members should not have been involved in other aspects of the project and should not have a vested interest in the outcome.

11.14.   Two methods of peer review should be used: participatory peer review and late-stage peer review. The participatory peer review should be conducted during the implementation of the project, allowing most of the comments of the reviewers to be resolved before the end of the project. The late-stage (follow-up) peer review should be conducted towards the end of the project. Participatory peer review decreases the likelihood that the results of the investigation of site characteristics and evaluation of radiation risks from the nuclear installation are found to be unsuitable at a later stage.

Appendix

APPLICATION OF A GRADED APPROACH TO MODELLING RADIONUCLIDE DISPERSION IN GROUNDWATER

A.1. This Appendix provides recommendations for determining the most appropriate level of complexity for modelling radionuclide dispersion in groundwater. Since different nuclear installations pose different levels of hazard, the suggested method uses a graded approach based on the level of hazard and the stage of reporting.

A.2. In addition to the level of hazard and the stage of reporting, several other factors may need to be taken into account in applying a graded approach. These factors include the complexity of the hydrogeological setting of the site; the type of the model; the type of solution provided by the mathematical model; the dimensionality of the model; the modes of flow and transport; the type of flow and transport domain; the availability, sources, reliability and representativeness of data needed for the selected model; the source and reliability of information on boundary conditions; and consideration of the processes that affect the transport and fate of radionuclides.

A.3. As described in para. 10.5, three or more levels of hazard categorization for nuclear installations may be defined. In this Appendix, three levels are assumed, based on the type and capacity of the nuclear installation.

A.4. Three reporting stages are assumed in this Appendix. Reporting Stage 1 relates to the site characterization stage, which involves a detailed study of the hydrogeological domain. Site specific data for hydrogeological conceptualization, characterization and modelling should be collected, evaluated and reported during this stage. Reporting Stage 2 relates to the construction and operation stages. For conformity of the analysis, this stage should include validation of the predictive model constructed for the site, using well established monitored observations of flow, hydraulic heads and concentrations. Validation is used here to mean a post-audit to assess the predictive accuracy of a site specific model based on long term monitoring data. Reporting Stage 3 relates to the closure of the installation. If there might be a new source term, its dispersion in groundwater should be simulated by a validated model.

A.5. The complexity of the hydrogeological configuration refers to the variety and contact relations of hydrostratigraphical units. Factors such as dual porosity, fracture and/or karst permeability, heterogeneity and anisotropy significantly complicate the hydrogeological setting and should be considered. The ease of construction of a representative hydrogeological conceptual model without oversimplification should also be considered during the evaluation stage.

A.6. Depending on the objective defined for the groundwater modelling study, the mathematical model should be selected to simulate different flow domains: saturated or a combination of saturated and unsaturated. The vadose zone has a significant role in the transport and fate of the radionuclides, in terms of retardation and attenuation. Therefore, it has a reductive effect on the contaminant transport. For a conservative prediction, this reductive effect should be omitted at the first stage of site selection and site evaluation. Simulating the flow and transport in the saturated flow domain could be sufficient to achieve the objectives. A simulation including the unsaturated flow domain is more complicated and needs data that are more difficult to acquire. A graded approach, as depicted in Figs 5–9, can be used to determine the conditions involving consideration of the vadose zone in the hydrogeological characterization.

A.7. The groundwater modelling can be achieved by using different techniques to solve the flow and transport equations. Partial differential equations simulating the groundwater flow and solute (radionuclide) transport are solved either analytically or numerically. The advantages and disadvantages of the different techniques are presented in detail in paras 6.26–6.34. It should be kept in mind that natural systems often do not exhibit configurations that closely match the geometries defined in specific analytical solutions. Therefore, analytical models should be used when the natural hydrogeological setting can be simplified with certain confidence to fit the assumptions of the analytical solution.

A.8. The objective and the level of hazard category may necessitate groundwater modelling in one, two or three dimensions. One dimensional models simulate flow and transport in the mean flow direction and should be used only for the low hazard category. For higher hazard categories, more detailed two or three dimensional models should be employed, particularly for the screening stage, to better capture spatial variability and potential impacts. Dimensionality should be selected on the basis of the objective, expected impact and level of hazard. The higher the hazard category and the higher the accuracy needed for the investigation, the more dimensions the model should have.

A.9. The selected model should then be run and calibrated for steady state (independent of time) and transient (time dependent) flow and transport modes. Calibration is achieved by the reproduction of observed heads and/or concentrations by the model. The model can be run only for steady flow and transport for the screening stage and/or for low hazard category installations. Transient flow and transport need to be simulated to make predictions. Therefore, the selected model (analytical or numerical) should be verified by checking the results against an independent set of data.

A.10. The application of a graded approach also involves the collection and use of different levels of data. For low hazard category installations and/or for the screening stage, data from literature, regional studies and information based on expert qualitative observations can be used if site specific data are not available. For higher levels of hazard category and at Reporting Stage 1, site specific representative data are needed. These data should include hydraulic parameters that represent all hydrostratigraphical units, the hydraulic head distribution at the flow domain and the hydraulic head, and concentration or fluxes at the boundaries of the hydrogeological domain.

A.11. A mathematical model to simulate the flow and transport processes should be selected on the basis of the level of hazard of the installation and the expected impact. The transport and fate of radionuclides in groundwater are primarily affected by advection, diffusion, sorption and radioactive decay, but processes such as dispersion should also be included if the results of groundwater modelling show that the site is not suitable for a nuclear installation. In some cases, transport models that include a reactive term may be selected.

A.12. There are several freeware and commercial computer codes that can be used in modelling studies. The model used should be verified to ensure that its numerical algorithm has been implemented correctly. In general, this is achieved by comparing the results of a numerical model with an analytical solution.

A.13. The following factors should be considered in selecting the model complexity:

(a)   Level of hazard;
(b)   Reporting stage;
(c)   Complexity of hydrogeological configuration;
(d)   Saturated or unsaturated media;
(e)   Dimensionality of the model;
(f)   Steady or transient flow mode;
(g)   Technique for solving the equations (i.e. analytical or numerical);
(h)   Source of data for parameters;
(i)   Source of information for boundary conditions;
(j)   Transport and fate processes.

A.14. The details of the application of a graded approach to groundwater modelling are illustrated in the flow charts in Figs 5–7 for Reporting Stage 1. Figures 8 and 9 illustrate the steps for Reporting Stages 2 and 3, respectively. The low and medium hazard categories are graded in three levels of complexity for the hydrogeological configuration of the site, while the high hazard category is graded in four levels.

A.15. The symbols and abbreviations used in the flow charts in Figs 5–9 are explained in Table 3.

A.16. A more detailed explanation for each step illustrated in Figs 5–9 is provided under paras A.17–A.19.

TABLE 3. SYMBOLS AND ABBREVIATIONS USED IN THE FLOW CHARTS IN FIGS 5–9

Symbol/abbreviation	Definition
Hazard category of nuclear installations	
LH_n	Low hazard category (complexity level n)
MH_n	Medium hazard category (complexity level n)
HH_n	High hazard category (complexity level n)
Stage of site evaluation	
Reporting Stage 1	Site characterization
Reporting Stage 2	Construction and operation
Reporting Stage 3	Decommissioning
Transfer mechanism	
v	Advective transfer constant
λ	Radioactive decay constant
α	Dispersive transfer constant
R_f	Sorption constant
r	First-order reaction constant

TABLE 3. SYMBOLS AND ABBREVIATIONS USED IN THE FLOW CHARTS IN FIGS 5–9 (cont.)

Symbol/abbreviation	Definition
Dimensionality and flow conditions	
1D/S	One dimensional and steady
2D/S	Two dimensional and steady
2D/ST	Two dimensional, steady and transient
3D/S	Three dimensional and steady
3D/ST	Three dimensional, steady and transient
Source of information	
Approx	Approximation based on site observations
BC	Source of boundary conditions
Lab	Data from tests at laboratory
Obs	From site observation
Pm	Source of parameter value
Reg	Data from literature in the close vicinity of the installation site
Site	Site specific representative data/field test
Model status	
C/S	Calibrated for steady state conditions
C/ST	Calibrated for steady state and transient conditions

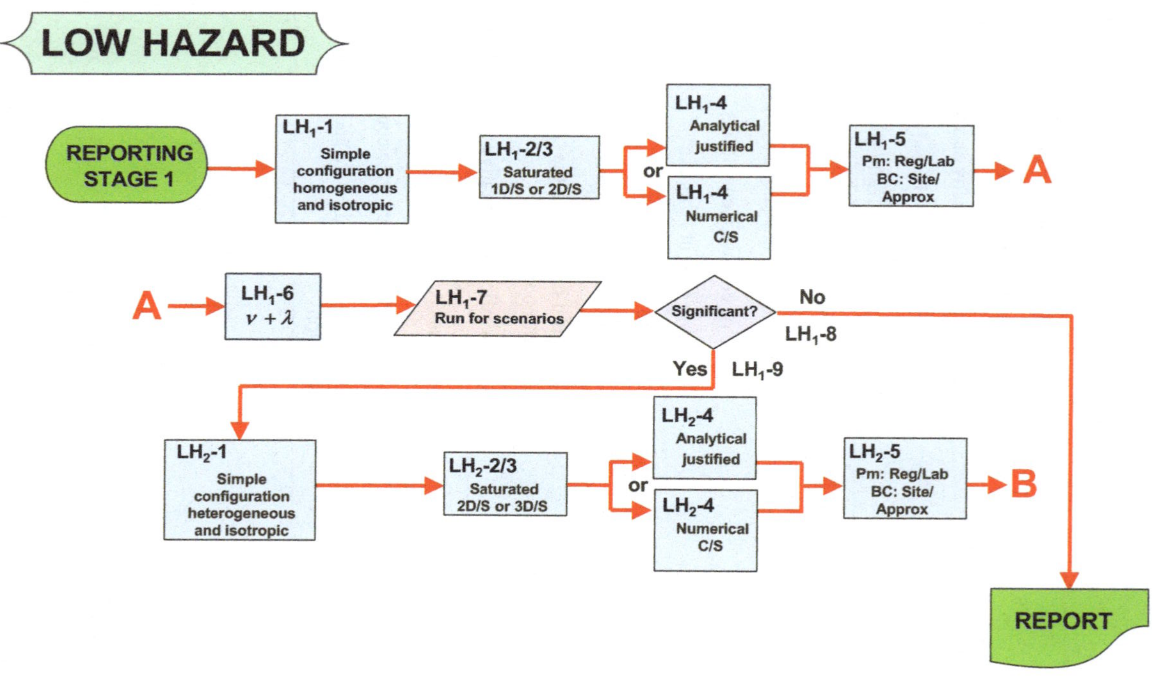

LOW HAZARD
REPORTING STAGE 1
LH$_1$-1
Simple configuration homogeneous and isotropic
LH$_1$-2/3
Saturated 1D/S or 2D/S
or
LH$_1$-4
Analytical justified
LH$_1$-4
Numerical C/S
LH$_1$-5
Pm: Reg/Lab
BC: Site/ Approx
A
A
LH$_1$-6
$\nu + \lambda$
LH$_1$-7
Run for scenarios
Significant?
No
LH$_1$-8
Yes
LH$_1$-9
LH$_2$-1
Simple configuration heterogeneous and isotropic
LH$_2$-2/3
Saturated 2D/S or 3D/S
or
LH$_2$-4
Analytical justified
LH$_2$-4
Numerical C/S
LH$_2$-5
Pm: Reg/Lab
BC: Site/ Approx
B
REPORT

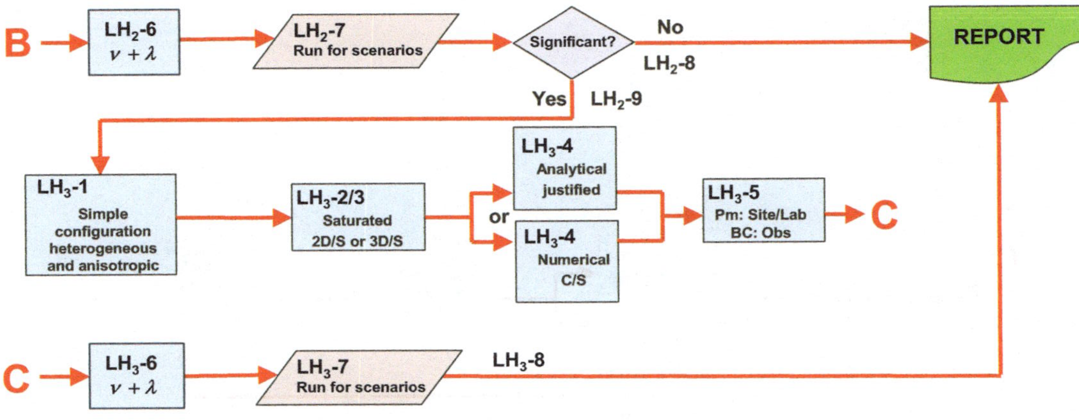

FIG. 5. Flow chart showing a graded approach to modelling radionuclide dispersion in groundwater in Reporting Stage 1 of site evaluation for low hazard category nuclear installations. (See Table 3 for definitions of the symbols and abbreviations.)

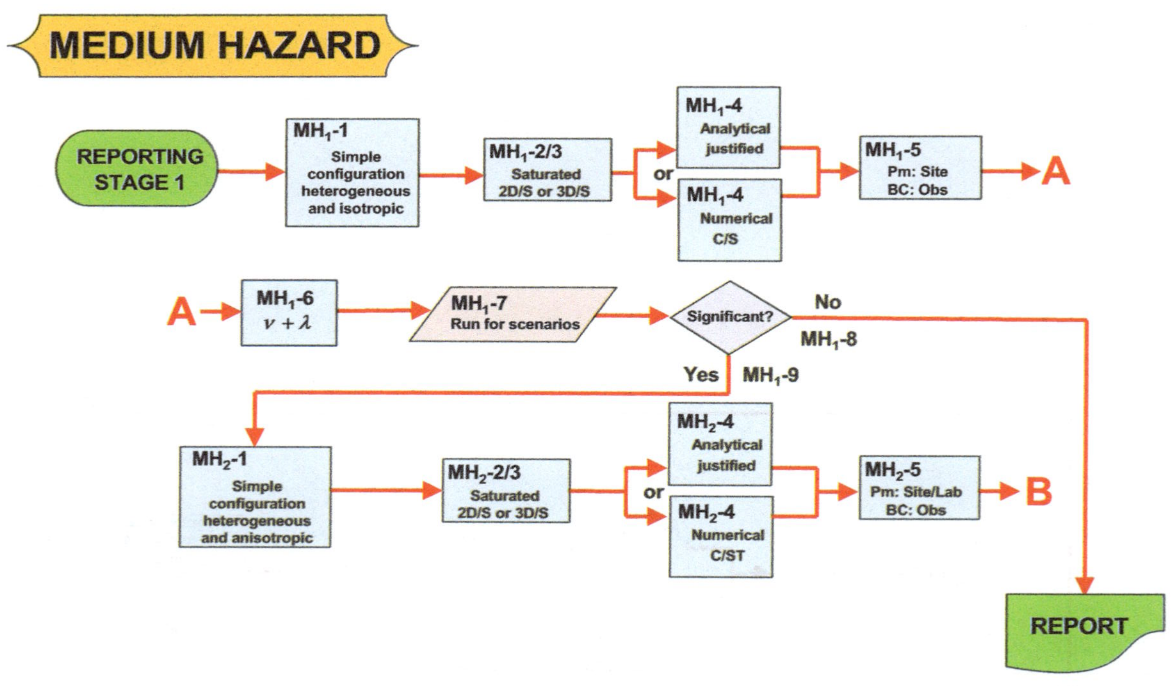
MEDIUM HAZARD
REPORTING STAGE 1
MH$_1$-1
Simple configuration heterogeneous and isotropic
MH$_1$-2/3
Saturated 2D/S or 3D/S
or
MH$_1$-4
Analytical justified
MH$_1$-4
Numerical C/S
MH$_1$-5
Pm: Site
BC: Obs
A
A
MH$_1$-6
$\nu + \lambda$
MH$_1$-7
Run for scenarios
Significant?
No
MH$_1$-8
Yes
MH$_1$-9
MH$_2$-1
Simple configuration heterogeneous and anisotropic
MH$_2$-2/3
Saturated 2D/S or 3D/S
or
MH$_2$-4
Analytical justified
MH$_2$-4
Numerical C/ST
MH$_2$-5
Pm: Site/Lab
BC: Obs
B
REPORT

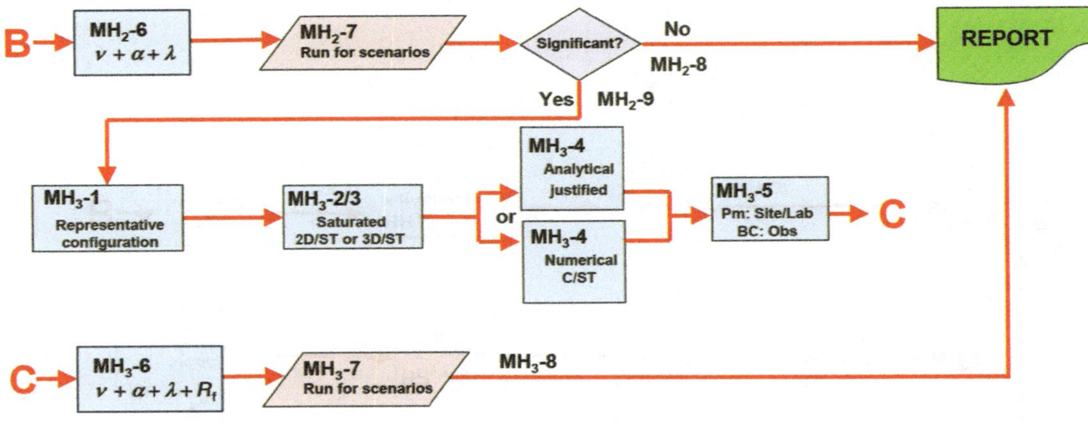

FIG. 6. Flow chart showing a graded approach to modelling radionuclide dispersion in groundwater in Reporting Stage 1 of site evaluation for medium hazard category nuclear installations. (See Table 3 for definitions of the symbols and abbreviations.)

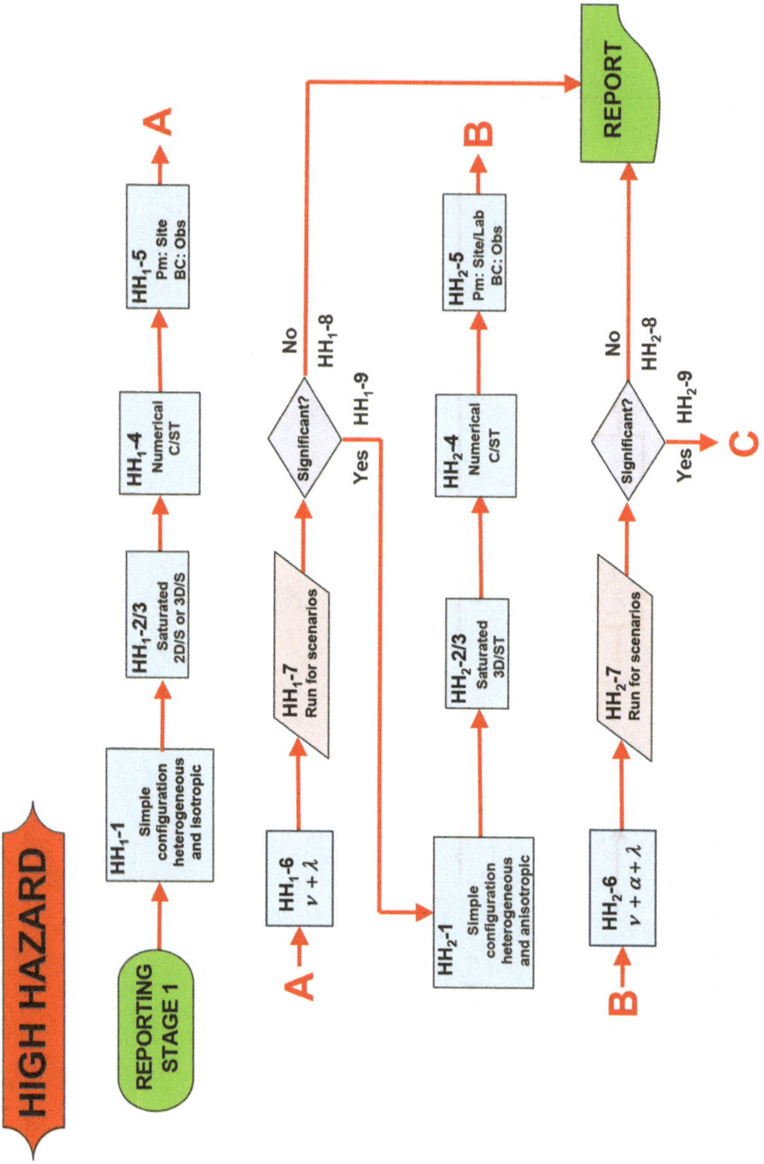

HIGH HAZARD
REPORTING STAGE 1
HH$_1$-1 Simple configuration heterogeneous and isotropic
HH$_1$-2/3 Saturated 2D/S or 3D/S
HH$_1$-4 Numerical C/ST
HH$_1$-5 Pm: Site BC: Obs
A
A
HH$_1$-6 $\nu + \lambda$
HH$_1$-7 Run for scenarios
Significant?
No HH$_1$-8
Yes HH$_1$-9
HH$_2$-1 Simple configuration heterogeneous and anisotropic
HH$_2$-2/3 Saturated 3D/ST
HH$_2$-4 Numerical C/ST
HH$_2$-5 Pm: Site/Lab BC: Obs
B
B
HH$_2$-6 $\nu + \alpha + \lambda$
HH$_2$-7 Run for scenarios
Significant?
No HH$_2$-8
Yes HH$_2$-9
C
REPORT

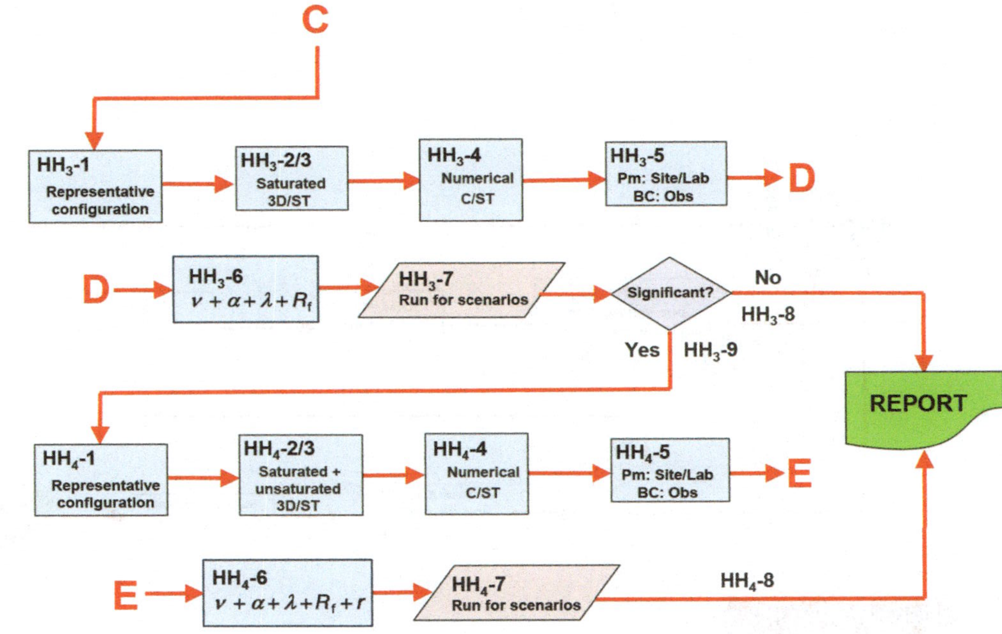

FIG. 7. *Flow chart showing a graded approach to modelling radionuclide dispersion in groundwater in Reporting Stage 1 of site evaluation for high hazard category nuclear installations. (See Table 3 for definitions of the symbols and abbreviations.)*

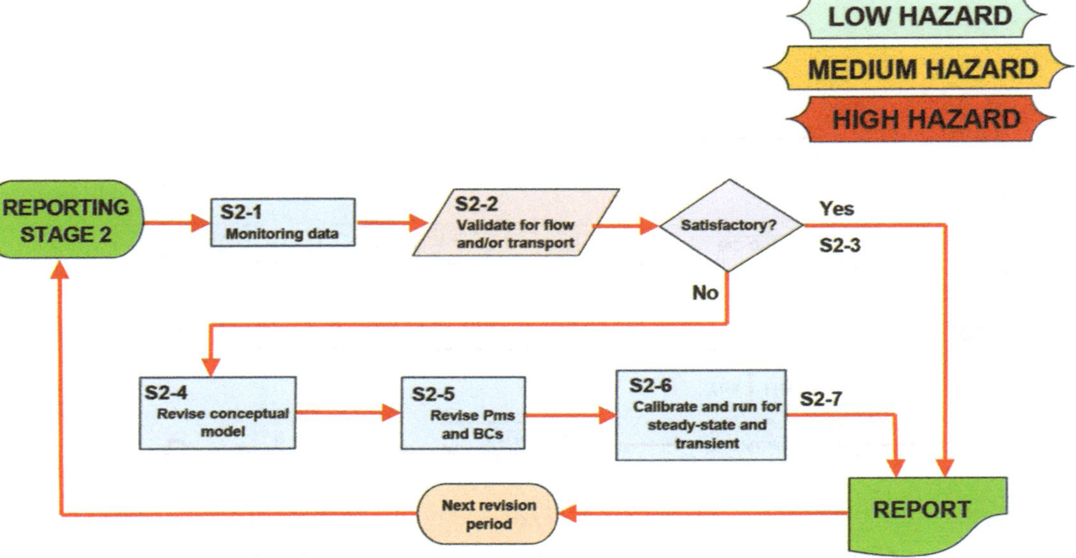

FIG. 8. Flow chart showing steps of modelling radionuclide dispersion in groundwater in Reporting Stage 2 of site evaluation for low, medium and high hazard category nuclear installations. (See Table 3 for definitions of the symbols and abbreviations.)

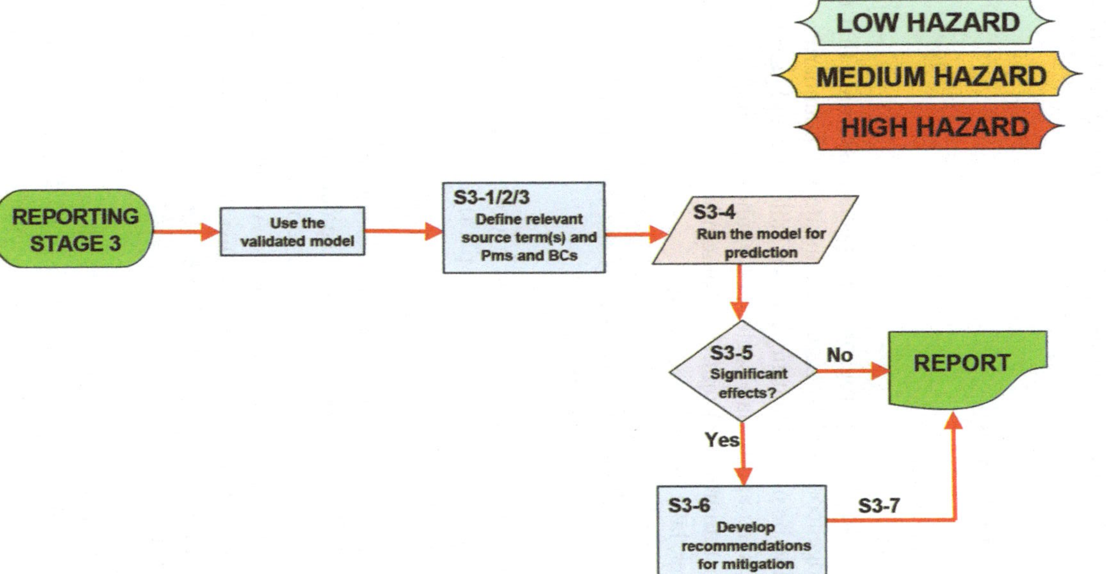

FIG. 9. *Flow chart showing steps of modelling radionuclide dispersion in groundwater in Reporting Stage 3 of site evaluation for low, medium and high hazard category nuclear installations. (See Table 3 for definitions of the symbols and abbreviations.)*

EXPLANATION OF EACH STEP FOR REPORTING STAGE 1

A.17. This section describes in detail the process depicted in Figs 5–7 for Reporting Stage 1, for different levels of hazard and for different levels of site complexity.

Reporting Stage 1: Low hazard

Low hazard — Complexity level 1

LH$_1$-1:   Conceptualize the hydrogeological system assuming that the aquifer is homogeneous and isotropic. Delineate the boundaries of the flow domain to be modelled using the available data and geological map of the site at an appropriate scale. The boundaries should be set either at a hydrostratigraphical unit (i.e. aquiclude) or a hydrological element (i.e. river, lake or measured head).

LH$_1$-2:   Consider only the saturated part of the flow domain and ignore the role of the unsaturated zone.

LH$_1$-3:   Construct a simple conceptual hydrogeological model in one dimension (along the main flow path) or two dimensions (cross-section and/or plane).

LH$_1$-4:   Select one or more analytical models for flow and solute transport and justify that any model used fits the constructed conceptual model in terms of geometry of the flow domain, processes considered, parameters and boundary conditions. Alternatively, select numerical model codes that simulate one dimensional or two dimensional flow and transport. The models should be calibrated and suitable for steady state conditions.

LH$_1$-5:   The hydraulic parameters (i.e. hydraulic conductivity, effective porosity) and transport parameters (e.g. dispersivity) may be taken from previous work at the site and vicinity and from laboratory tests. The hydraulic and concentration boundary conditions can be defined or estimated by expert approximation and observations at the site.

LH$_1$-6:   Choose models that simulate the advective transport and radioactive decay of specific radionuclides expected to be released from the installation.

LH$_1$-7:   Run the calibrated model for scenarios.

LH$_1$-8:   Document whether contaminant concentrations are acceptable.

LH$_1$-9:   If the results show that the concentrations of contaminants are significant, move to a higher complexity level of investigation and modelling (LH$_2$).

Low hazard — Complexity level 2

LH$_2$-1:   Conceptualize the hydrogeological system considering the heterogeneity of the flow domain. Assume the flow domain is isotropic. Delineate the boundaries of the flow domain to be modelled using the available data and geological map of the site at an appropriate scale. The boundaries should be set either at a hydrostratigraphical unit (i.e. aquiclude) or a hydrological element (i.e. river, lake or measured head).

LH$_2$-2:   Consider only the saturated part of the flow domain and ignore the role of the unsaturated zone.

LH$_2$-3:   Construct a simple conceptual hydrogeological model in two dimensions (cross-section and/or plane) or three dimensions (e.g. block or fence diagram).

LH$_2$-4:   Select one or more analytical model for flow and solute transport and justify that any model used fits the constructed conceptual model in terms of geometry of the flow domain, processes considered, parameters and boundary conditions. Alternatively, select numerical model codes that simulate one dimensional and/or two dimensional flow and transport. The models should be calibrated and suitable for steady state conditions.

LH$_2$-5:   The hydraulic parameters (i.e. hydraulic conductivity, effective porosity) may be taken from previous work at the site and vicinity and from laboratory tests. The dispersivity (or hydrodynamic dispersion coefficient) can be taken from literature for similar types of hydrogeological units. The hydraulic and concentration boundary conditions can be defined or estimated by expert approximation and observations at the site.

LH$_2$-6:   Choose models that simulate the advective and dispersive transport and radioactive decay of specific radionuclides expected to be released from the installation.

LH$_2$-7:   Run the calibrated model for scenarios.

LH$_2$-8:   Document whether contaminant concentrations are acceptable.

LH$_2$-9:   If the results show that the concentrations of contaminants are significant, move to a higher complexity level of investigation and modelling (LH$_3$).

Low hazard — Complexity level 3

LH$_3$-1:   Conceptualize the hydrogeological system considering the heterogeneity and anisotropy of the flow domain. When site data are lacking, the anisotropy can be assumed such that horizontal hydraulic conductivity is ten times higher than vertical hydraulic conductivity. Delineate the boundaries of the flow domain to be modelled using the available data and geological map of the site at an appropriate scale. The boundaries should be set either at a hydrostratigraphical unit (i.e. aquiclude) or a hydrological element (i.e. river, lake or measured head).

LH$_3$-2:   Consider only the saturated part of the flow domain and ignore the role of the unsaturated zone.

LH$_3$-3:   Construct a simple conceptual hydrogeological model in two dimensions (cross-section and/or plane) or three dimensions (e.g. block or fence diagram).

LH$_3$-4:   Select one or more analytical models for flow and solute transport and justify that any model used fits the constructed conceptual model in terms of geometry of the flow domain, processes considered, parameters and boundary conditions. Alternatively, select numerical model codes that simulate two dimensional or three dimensional flow and transport. The models should be calibrated and suitable for steady state conditions.

LH$_3$-5:   The hydraulic parameters (i.e. hydraulic conductivity, effective porosity) may be taken from previous work at the site and from laboratory tests. The dispersivity (or hydrodynamic dispersion coefficient) should be obtained from laboratory or in situ tests. The hydraulic and concentration boundary conditions can be defined on the basis of observations at the site.

LH$_3$-6:   Choose models that simulate the advective and dispersive transport and radioactive decay of specific radionuclides expected to be released from the installation.

LH$_3$-7:   Run the calibrated model for scenarios.

LH$_3$-8:   Document the results.

Reporting Stage 1: Medium hazard

Medium hazard — Complexity level 1

MH$_1$-1:   Conceptualize the hydrogeological system considering the heterogeneity of the medium and assuming that the aquifer is isotropic. Delineate the boundaries of the flow domain to be modelled using the available data and geological map of the site at an appropriate scale. The boundaries should be set either at a hydrostratigraphical unit (i.e. aquiclude) or a hydrological element (i.e. river, lake or measured head).

MH$_1$-2:   Consider only the saturated part of the flow domain and ignore the role of the unsaturated zone.

MH$_1$-3: Construct a simple conceptual hydrogeological model in two dimensions (cross-section and/or plane) or three dimensions (e.g. block or fence diagram).

MH$_1$-4: Select one or more analytical models for flow and solute transport and justify that any model used fits the constructed conceptual model, in terms of geometry of the flow domain, processes considered, parameters and boundary conditions. Alternatively, select numerical model codes that simulate two dimensional or three dimensional flow and transport. The models should be calibrated and suitable for steady state conditions.

MH$_1$-5: The hydraulic characteristics (e.g. transmissivity, storativity) should be obtained from tests at the site and previous work (if any) at the site. The hydraulic and concentration boundary conditions should be defined on the basis of observations at the site.

MH$_1$-6: Choose models that simulate the advective transport and radioactive decay of specific radionuclides expected to be released from the installation.

MH$_1$-7: Run the calibrated model for scenarios.

MH$_1$-8: Document whether contaminant concentrations are acceptable.

MH$_1$-9: If the results show that the concentrations of contaminants are significant, move to a higher complexity level of investigation and modelling (MH$_2$).

Medium hazard — Complexity level 2

MH$_2$-1: Conceptualize the hydrogeological system considering the heterogeneity and anisotropy of the flow domain. When site data are lacking, the anisotropy can be assumed such that horizontal hydraulic conductivity is ten times higher than vertical hydraulic conductivity. Delineate the boundaries of the flow domain to be modelled using the available data and geological map of the site at an appropriate scale. The boundaries should be set either at a hydrostratigraphical unit (i.e. aquiclude) or a hydrological element (i.e. river, lake or measured head).

MH$_2$-2:   Consider only the saturated part of the flow domain and ignore the role of the unsaturated zone.

MH$_2$-3:   Construct a simple conceptual hydrogeological model in two dimensions (cross-section and/or plane) or three dimensions (e.g. block or fence diagram). Oversimplification of the real conditions should be avoided.

MH$_2$-4:   Select analytical models for flow and solute transport and justify that they fit the constructed conceptual model in terms of geometry of the flow domain, processes considered, parameters and boundary conditions. Alternatively, select numerical model codes that simulate two dimensional or three dimensional flow and transport. The models should be calibrated and suitable for steady state and transient conditions.

MH$_2$-5:   The hydraulic characteristics (e.g. transmissivity, storativity) should be obtained from tests at the site and previous work (if any) at the site. The dispersivity (or hydrodynamic dispersion coefficient) should be obtained from the laboratory or in situ tests. The hydraulic and concentration boundary conditions should be defined on the basis of observations at the site.

MH$_2$-6:   Choose models that simulate the advective and dispersive transport and radioactive decay of specific radionuclides expected to be released from the installation.

MH$_2$-7:   Run the calibrated model for scenarios.

MH$_2$-8:   Document whether contaminant concentrations are acceptable.

MH$_2$-9:   If the results show that the concentrations of contaminants are significant, move to a higher complexity level of investigation and modelling (MH$_3$).

MH$_3$-1: Conceptualize the hydrogeological system considering the heterogeneity and anisotropy of the flow domain. Special care should be taken to avoid oversimplification to ensure that the constructed conceptual hydrogeological model represents the real conditions. Delineate the boundaries of the flow domain to be modelled using the detailed geological mapping of the site at an appropriate scale. The boundaries should be set either at a hydrostratigraphical unit (i.e. aquiclude) or a hydrological element (i.e. river, lake or measured head). The boundary type should be defined and quantified from observations using appropriate techniques (e.g. piezometers) at the site.

MH$_3$-2: Consider only the saturated part of the flow domain and ignore the role of the unsaturated zone.

MH$_3$-3: Construct a detailed and representative conceptual hydrogeological model in two dimensions (cross-section and/or plane) or three dimensions (e.g. block or fence diagram). Oversimplification of the real conditions should be avoided.

MH$_3$-4: Use analytical models for flow and solute transport and justify that they fit the constructed conceptual model in terms of geometry of the flow domain, processes considered, parameters and boundary conditions. Alternatively, select numerical model codes that simulate two dimensional or three dimensional flow and transport. The models should be calibrated and suitable for steady state and/or transient conditions.

MH$_3$-5: The hydraulic characteristics (e.g. transmissivity, storativity) should be obtained from tests at the site and previous work (if any) at the site and/or from laboratory tests. The dispersivity (or hydrodynamic dispersion coefficient) and distribution coefficient should be obtained from the laboratory or in situ tests. The hydraulic and concentration boundary conditions can be defined on the basis of observations at the site.

MH$_3$-6: Choose models that simulate the advective, dispersive and sorptive (with retardation) transport and radioactive decay of specific radionuclides expected to be released from the installation.

MH$_3$-7:   Run the calibrated model for scenarios.

MH$_3$-8:   Document the results.

Reporting Stage 1: High hazard

High hazard — Complexity level 1

HH$_1$-1:   Conceptualize the hydrogeological system considering the heterogeneity of the medium and assuming that the aquifer is isotropic. Delineate the boundaries of the flow domain to be modelled using the available data and geological map of the site at an appropriate scale. The boundaries should be set either at a hydrostratigraphical unit (i.e. aquiclude) or a hydrological element (i.e. river, lake or measured head).

HH$_1$-2:   Consider only the saturated part of the flow domain and ignore the role of the unsaturated zone.

HH$_1$-3:   Construct a simple conceptual hydrogeological model in two dimensions (cross-section and/or plane) or three dimensions (e.g. block or fence diagram). Oversimplification of the real conditions should be avoided.

HH$_1$-4:   Select numerical model codes that simulate two dimensional or three dimensional flow and transport. The models should be calibrated and suitable for steady state and transient conditions.

HH$_1$-5:   The hydraulic characteristics (e.g. transmissivity, storativity) should be obtained from tests at the site and previous work (if any) at the site. The hydraulic and concentration boundary conditions should be defined on the basis of observations at the site.

HH$_1$-6:   Choose models that simulate the advective transport and radioactive decay of specific radionuclides expected to be released from the installation.

HH$_1$-7:   Run the calibrated model for scenarios.

HH$_1$-8:   Document whether contaminant concentrations are acceptable.

HH_1-9:   If the results show that the concentrations of contaminants are significant, move to a higher complexity level of investigation and modelling (HH_2)

High hazard — Complexity level 2

HH_2-1:   Conceptualize the hydrogeological system considering the heterogeneity and anisotropy of the flow domain. When site data are lacking, the anisotropy can be assumed such that horizontal hydraulic conductivity is ten times higher than vertical hydraulic conductivity. Delineate the boundaries of the flow domain to be modelled using the available data and geological map of the site at an appropriate scale. The boundaries should be set either at a hydrostratigraphical unit (i.e. aquiclude) or a hydrological element (i.e. river, lake or measured head).

HH_2-2:   Consider only the saturated part of the flow domain and ignore the role of the unsaturated zone.

HH_2-3:   Construct a conceptual hydrogeological model in three dimensions (e.g. block or fence diagram). Oversimplification of the real conditions should be avoided.

HH_2-4:   Select numerical model codes that simulate three dimensional flow and transport. The models should be calibrated and suitable for steady state and transient conditions.

HH_2-5:   The hydraulic characteristics (e.g. transmissivity, storativity) should be obtained from tests at the site and previous work (if any) at the site. The dispersivity (or hydrodynamic dispersion coefficient) should be obtained from the laboratory or in situ tests. The hydraulic and concentration boundary conditions should be defined on the basis of observations at the site.

HH_2-6:   Choose models that simulate the advective and dispersive transport and radioactive decay of specific radionuclides expected to be released from the installation.

HH_2-7:   Run the calibrated model for scenarios.

HH$_2$-8:   Document whether contaminant concentrations are acceptable.

HH$_2$-9:   If the results show that the concentrations of contaminants are significant, move to a higher complexity level of investigation and modelling (HH$_3$).

High hazard — Complexity level 3

HH$_3$-1:   Conceptualize the hydrogeological system considering the heterogeneity and anisotropy of the flow domain. Special care should be taken to avoid oversimplification to ensure that the constructed conceptual hydrogeological model represents the real conditions. Delineate the boundaries of the flow domain to be modelled using the detailed geological mapping of the site at an appropriate scale. The boundaries should be set either at a hydrostratigraphical unit (i.e. aquiclude) or a hydrological element (i.e. river, lake or measured head). The boundary type should be defined and quantified from observations using appropriate techniques (e.g. piezometers) at the site.

HH$_3$-2:   Consider only the saturated part of the flow domain and ignore the role of the unsaturated zone.

HH$_3$-3:   Construct a detailed and representative conceptual hydrogeological model in three dimensions (e.g. block or fence diagram).

HH$_3$-4:   Select numerical model codes that simulate three dimensional flow and transport. The model should be calibrated and suitable for steady state and transient conditions.

HH$_3$-5:   The hydraulic characteristics (e.g. transmissivity, storativity) should be obtained from tests at the site and previous work (if any) at the site and/or from laboratory tests. The dispersivity (or hydrodynamic dispersion coefficient) and distribution coefficient should be obtained from the laboratory and/or in situ tests. The hydraulic and concentration boundary conditions can be defined on the basis of observations at the site.

HH$_3$-6:   Choose models that simulate the advective, dispersive and sorptive (with retardation) transport and radioactive decay of specific radionuclides expected to be released from the installation.

HH$_3$-7:   Run the calibrated model for scenarios.

HH$_3$-8:   Document whether contaminant concentrations are acceptable.

HH$_3$-9:   If the results show that the concentrations of contaminants are significant, go to a higher complexity level of investigation and modelling (HH$_4$).

High hazard — Complexity level 4

HH$_4$-1:   Conceptualize the hydrogeological system considering the heterogeneity and anisotropy of the flow domain. Special care should be taken to avoid oversimplification to ensure that the constructed conceptual hydrogeological model represents the real conditions. Delineate the boundaries of the flow domain to be modelled using the detailed geological mapping of the site at an appropriate scale. The boundaries should be set either at a hydrostratigraphical unit (i.e. aquiclude) or a hydrological element (i.e. river, lake or measured head). The boundary type should be defined and quantified from observations using appropriate techniques (e.g. piezometers) at the site.

HH$_4$-2:   Consider the saturated and the unsaturated zone of the flow domain.

HH$_4$-3:   Construct a detailed and representative conceptual hydrogeological model in three dimensions (e.g. block or fence diagram) including the unsaturated zone.

HH$_4$-4:   Select numerical model codes that simulate three dimensional saturated or unsaturated flow and transport. The model(s) should be calibrated and suitable for steady state and transient conditions.

HH$_4$-5:   The hydraulic characteristics (e.g. transmissivity, storativity) should be obtained from tests at the site. Data from previous work (if any) at the site and/or from laboratory tests may also be used if consistent. The dispersivity (or hydrodynamic dispersion coefficient) and distribution coefficient should be obtained from in situ tests and laboratory tests. Reaction rate coefficients can be taken from literature or from laboratory tests. The hydraulic and concentration boundary conditions can be defined on the basis of observations at the site.

HH$_4$-6:   Choose models that simulate the advective, dispersive, sorptive (with retardation) and reactive (if found significant) transport and radioactive decay of specific radionuclides expected to be released from the installation.

HH$_4$-7:   Run the calibrated model for scenarios.

HH$_4$-8:   Document the results.

EXPLANATION OF STEPS FOR REPORTING STAGE 2

A.18. This section describes the process depicted in Fig. 8 for Reporting Stage 2. This reporting stage includes the periodic validation of the models using the data obtained from the monitoring network and the revision of the conceptual model if necessary.

Reporting Stage 2: All hazard categories

S2-1:   Collect the flow, head and quality data from the monitoring network at the site.

S2-2:   Compare the observed data with those calculated by the models run to predict the future conditions.

S2-3:   Document whether the model predictions are satisfactory.

S2-4:   If the predictions are poor, revise the conceptual model.

S2-5:   Check and revise the parameters and boundary conditions.

S2-6:   Calibrate and run the model for steady state and transient conditions.

S2-7:   Report and continue monitoring until the next period of validation.

EXPLANATION OF STEPS FOR REPORTING STAGE 3

A.19. This section describes the process depicted in Fig. 9 for Reporting Stage 3. This reporting stage includes the use of the validated model to predict the impacts during decommissioning, which may assist in developing suggestions for mitigating adverse effects, if any.

Reporting Stage 3: All hazard categories

S3-1:   Identify and define any new source terms created by decommissioning activities.

S3-2:   Identify and define any new boundary conditions created by decommissioning activities.

S3-3:   Identify and quantify any alterations of hydraulic parameters.

S3-4:   Run the validated model.

S3-5:   Identify any adverse effects on water resources.

S3-6:   Develop potential mitigation measures for any adverse effects and model their performance.

S3-7:   Report the results.

REFERENCES

[1] INTERNATIONAL ATOMIC ENERGY AGENCY, Site Evaluation for Nuclear Installations, IAEA Safety Standards Series No. SSR-1, IAEA, Vienna (2019).

[2] INTERNATIONAL ATOMIC ENERGY AGENCY, Safety of Research Reactors, IAEA Safety Standards Series No. SSR-3, IAEA, Vienna (2016).

[3] INTERNATIONAL ATOMIC ENERGY AGENCY, Safety of Nuclear Fuel Cycle Facilities, IAEA Safety Standards Series No. SSR-4, IAEA, Vienna (2017).

[4] EUROPEAN COMMISSION, FOOD AND AGRICULTURE ORGANIZATION OF THE UNITED NATIONS, INTERNATIONAL ATOMIC ENERGY AGENCY, INTERNATIONAL LABOUR ORGANIZATION, OECD NUCLEAR ENERGY AGENCY, PAN AMERICAN HEALTH ORGANIZATION, UNITED NATIONS ENVIRONMENT PROGRAMME, WORLD HEALTH ORGANIZATION, Radiation Protection and Safety of Radiation Sources: International Basic Safety Standards, IAEA Safety Standards Series No. GSR Part 3, IAEA, Vienna (2014),
https://doi.org/10.61092/iaea.u2pu-60vm

[5] FOOD AND AGRICULTURE ORGANIZATION OF THE UNITED NATIONS, INTERNATIONAL ATOMIC ENERGY AGENCY, INTERNATIONAL CIVIL AVIATION ORGANIZATION, INTERNATIONAL LABOUR ORGANIZATION, INTERNATIONAL MARITIME ORGANIZATION, INTERPOL, OECD NUCLEAR ENERGY AGENCY, PAN AMERICAN HEALTH ORGANIZATION, PREPARATORY COMMISSION FOR THE COMPREHENSIVE NUCLEAR-TEST-BAN TREATY ORGANIZATION, UNITED NATIONS ENVIRONMENT PROGRAMME, UNITED NATIONS OFFICE FOR THE COORDINATION OF HUMANITARIAN AFFAIRS, WORLD HEALTH ORGANIZATION, WORLD METEOROLOGICAL ORGANIZATION, Preparedness and Response for a Nuclear or Radiological Emergency, IAEA Safety Standards Series No. GSR Part 7, IAEA, Vienna (2015),
https://doi.org/10.61092/iaea.3dbe-055p

[6] INTERNATIONAL ATOMIC ENERGY AGENCY, IAEA Nuclear Safety and Security Glossary: Terminology Used in Nuclear Safety, Nuclear Security, Radiation Protection and Emergency Preparedness and Response, 2022 (Interim) Edition, IAEA, Vienna (2022),
https://doi.org/10.61092/iaea.rrxi-t56z

[7] INTERNATIONAL ATOMIC ENERGY AGENCY, WORLD METEOROLOGICAL ORGANIZATION, Meteorological and Hydrological Hazards in Site Evaluation for Nuclear Installations, IAEA Safety Standards Series No. SSG-18, IAEA, Vienna (2011). (A revision of this publication is in preparation.)

[8] INTERNATIONAL ATOMIC ENERGY AGENCY, Seismic Hazards in Site Evaluation for Nuclear Installations, IAEA Safety Standards Series No. SSG-9 (Rev. 1), IAEA, Vienna (2022).

[9] INTERNATIONAL ATOMIC ENERGY AGENCY, Geotechnical Aspects in the Siting and Design of Nuclear Installations, IAEA Safety Standards Series No. SSG-93, IAEA, Vienna (in press).

[10] INTERNATIONAL ATOMIC ENERGY AGENCY, Volcanic Hazards in Site Evaluation for Nuclear Installations, IAEA Safety Standards Series No. SSG-21, IAEA, Vienna (2012).

[11] INTERNATIONAL ATOMIC ENERGY AGENCY, Hazards Associated with Human Induced External Events in Site Evaluation for Nuclear Installations, IAEA Safety Standards Series No. SSG-79, IAEA, Vienna (2023).

[12] INTERNATIONAL ATOMIC ENERGY AGENCY, Identification and Categorization of Sabotage Targets, and Identification of Vital Areas at Nuclear Facilities, IAEA Nuclear Security Series No. 48-T, IAEA, Vienna (2025),
https://doi.org/10.61092/iaea.74e6-e2yc

[13] INTERNATIONAL ATOMIC ENERGY AGENCY, UNITED NATIONS ENVIRONMENT PROGRAMME, Prospective Radiological Environmental Impact Assessment for Facilities and Activities, IAEA Safety Standards Series No. GSG-10, IAEA, Vienna (2018).

[14] INTERNATIONAL ATOMIC ENERGY AGENCY, Monitoring for Protection of the Public and the Environment, IAEA Safety Standards Series No. GSG-19, IAEA, Vienna (2025),
https://doi.org/10.61092/iaea.i7sy-xqd1

[15] INTERNATIONAL NUCLEAR SAFETY ADVISORY GROUP, Potential Exposure in Nuclear Safety, INSAG Series No. 9, IAEA, Vienna (1995).

[16] INTERNATIONAL ATOMIC ENERGY AGENCY, Deterministic Safety Analysis for Nuclear Power Plants, IAEA Safety Standards Series No. SSG-2 (Rev. 1), IAEA, Vienna (2019).

[17] UNITED NATIONS, Ionizing Radiation: Sources and Biological Effects (Report to the General Assembly), Scientific Committee on the Effects of Atomic Radiation (UNSCEAR), Supplement No. 46, UN, New York (2012).

[18] INTERNATIONAL ATOMIC ENERGY AGENCY, Development and Application of Level 2 Probabilistic Safety Assessment for Nuclear Power Plants, IAEA Safety Standards Series No. SSG-4 (Rev. 1), IAEA, Vienna (2025),
https://doi.org/10.61092/iaea.buld-d6ia

[19] INTERNATIONAL COMMISSION ON RADIOLOGICAL PROTECTION, Assessing Dose of the Representative Person for the Purpose of the Radiation Protection of the Public and The Optimisation of Radiological Protection: Broadening the Process, ICRP Publication 101, Elsevier, Oxford (2006).

[20] INTERNATIONAL ATOMIC ENERGY AGENCY, UNITED NATIONS ENVIRONMENT PROGRAMME, Regulatory Control of Radioactive Discharges to the Environment, IAEA Safety Standards Series No. GSG-9, IAEA, Vienna (2018).

[21] WORLD METEOROLOGICAL ORGANIZATION, Guidelines on Meteorological and Hydrological Aspects of Siting and Operation of Nuclear Power Plants, 2020 edn, WMO-No. 550, WMO, Geneva (2020).

[22] WORLD METEOROLOGICAL ORGANIZATION, Guide to Instruments and Methods of Observation, Vol. 1 – Measurement of Meteorological Variables, 2023 edn, WMO-No. 8, WMO, Geneva (2023).

[23] INTERNATIONAL ATOMIC ENERGY AGENCY, Development and Application of Level 1 Probabilistic Safety Assessment for Nuclear Power Plants, IAEA Safety Standards Series No. SSG-3 (Rev. 1), IAEA, Vienna (2024),
https://doi.org/10.61092/iaea.3ezv-lp49

[24] INTERNATIONAL ATOMIC ENERGY AGENCY, Radiation Protection Aspects of Design for Nuclear Power Plants, IAEA Safety Standards Series No. SSG-90, IAEA, Vienna (2024),
https://doi.org/10.61092/iaea.jc6f-diaa

[25] INTERNATIONAL COMMISSION ON RADIOLOGICAL PROTECTION, Occupational Intakes of Radionuclides: Part 1, ICRP Publication 130, SAGE, Thousand Oaks, CA (2015),
https://doi.org/10.1177/0146645315577539

[26] INTERNATIONAL COMMISSION ON RADIOLOGICAL PROTECTION, Compendium of Dose Coefficients based on ICRP Publication 60, ICRP Publication 119, Elsevier, Oxford (2012).

[27] INTERNATIONAL COMMISSION ON RADIOLOGICAL PROTECTION, Dose Coefficients for External Exposures to Environmental Sources, ICRP Publication 144, SAGE, Thousand Oaks, CA (2020),
https://doi.org/10.1177/0146645320906277

[28] INTERNATIONAL COMMISSION ON RADIOLOGICAL PROTECTION, Environmental Protection: The Concept and Use of Reference Animals and Plants, ICRP Publication 108, Elsevier, Oxford (2008).

[29] INTERNATIONAL ATOMIC ENERGY AGENCY, UNITED NATIONS ENVIRONMENT PROGRAMME, Radiation Protection of the Public and the Environment, IAEA Safety Standards Series No. GSG-8, IAEA, Vienna (2018).

[30] INTERNATIONAL COMMISSION ON RADIOLOGICAL PROTECTION, Protection from Potential Exposure: A Conceptual Framework, ICRP Publication 64 (1993),
https://doi.org/10.1016/0146-6453(93)90066-H

[31] INTERNATIONAL COMMISSION ON RADIOLOGICAL PROTECTION, The 2007 Recommendations of the International Commission on Radiological Protection, ICRP Publication 103, Elsevier, Oxford (2007).

[32] INTERNATIONAL COMMISSION ON RADIOLOGICAL PROTECTION, Environmental Protection: Transfer Parameters for Reference Animals and Plants, ICRP Publication 114, Elsevier, Oxford (2009),
https://doi.org/10.1016/j.icrp.2011.08.009

[33] INTERNATIONAL COMMISSION ON RADIOLOGICAL PROTECTION, Protection of the Environment Under Different Exposure Situations, ICRP Publication 124, SAGE, Thousand Oaks, CA (2014),
https://doi.org/10.1177/0146645313497456

[34] INTERNATIONAL COMMISSION ON RADIOLOGICAL PROTECTION, Dose Coefficients for Non-human Biota Environmentally Exposed to Radiation, ICRP Publication 136, SAGE, Thousand Oaks, CA (2017),
https://doi.org/10.1177/0146645317728022

[35] INTERNATIONAL ATOMIC ENERGY AGENCY, Guidelines on Soil and Vegetation Sampling for Radiological Monitoring, Technical Reports Series No. 486, IAEA, Vienna (2019).

[36] INTERNATIONAL ATOMIC ENERGY AGENCY, Programmes and Systems for Source and Environmental Radiation Monitoring, Safety Reports Series No. 64, IAEA, Vienna (2010).

[37] INTERNATIONAL ATOMIC ENERGY AGENCY, INTERNATIONAL LABOUR OFFICE, Occupational Radiation Protection, IAEA Safety Standards Series No. GSG-7, IAEA, Vienna (2018).

[38] INTERNATIONAL ATOMIC ENERGY AGENCY, Optimization of Radiation Protection in the Control of Occupational Exposure, Safety Reports Series No. 21, IAEA, Vienna (2002).

[39] INTERNATIONAL ATOMIC ENERGY AGENCY, Decommissioning of Facilities, IAEA Safety Standards Series No. GSR Part 6, IAEA, Vienna (2014).

[40] INTERNATIONAL ATOMIC ENERGY AGENCY, Decommissioning of Nuclear Power Plants, Research Reactors and Other Nuclear Fuel Cycle Facilities, IAEA Safety Standards Series No. SSG-47, IAEA, Vienna (2018).

[41] INTERNATIONAL ATOMIC ENERGY AGENCY, Policies and Strategies for the Decommissioning of Nuclear and Radiological Facilities, IAEA Nuclear Energy Series No. NW-G-2.1, IAEA, Vienna (2011).

[42] FOOD AND AGRICULTURE ORGANIZATION OF THE UNITED NATIONS, INTERNATIONAL ATOMIC ENERGY AGENCY, INTERNATIONAL LABOUR OFFICE, PAN AMERICAN HEALTH ORGANIZATION, UNITED NATIONS OFFICE FOR THE COORDINATION OF HUMANITARIAN AFFAIRS, WORLD HEALTH ORGANIZATION, Arrangements for Preparedness for a Nuclear or Radiological Emergency, IAEA Safety Standards Series No. GS-G-2.1, IAEA, Vienna (2007). (A revision of this publication is in preparation.)

[43] INTERNATIONAL ATOMIC ENERGY AGENCY, Leadership and Management for Safety, IAEA Safety Standards Series No. GSR Part 2, IAEA, Vienna (2016), https://doi.org/10.61092/iaea.cq1k-j5z3

CONTRIBUTORS TO DRAFTING AND REVIEW

Altinyollar, A.	International Atomic Energy Agency
Avci, H.	Consultant, United States of America
Contri, P.	International Atomic Energy Agency
Ekmekci, M.	Hacettepe University, Türkiye
Harman, N.	Consultant, United Kingdom
Kaeriyama, H.	Japan Fisheries Research and Education Agency, Japan
McDuffie, S.	International Atomic Energy Agency
Telleria, D.	International Atomic Energy Agency

CONTACT IAEA PUBLISHING

Feedback on IAEA publications may be given via the on-line form available at:
www.iaea.org/publications/feedback

This form may also be used to report safety issues or environmental queries concerning IAEA publications.

Alternatively, contact IAEA Publishing:

Publishing Section
International Atomic Energy Agency
Vienna International Centre, PO Box 100, 1400 Vienna, Austria
Telephone: +43 1 2600 22529 or 22530
Email: sales.publications@iaea.org
www.iaea.org/publications

Priced and unpriced IAEA publications may be ordered directly from the IAEA.

ORDERING LOCALLY

Priced IAEA publications may be purchased from regional distributors and from major local booksellers.

Printed and bound by CPI Group (UK) Ltd, Croydon, CR0 4YY

06/07/2026

02160596-0004